ADVANCED CALCULUS I

Staff of Research and Education Association,
Dr. M. Fogiel, Director

This book covers the usual course outline of Advanced Calculus I. For more advanced topics, see *"THE ESSENTIALS OF ADVANCED CALCULUS II"*.

Research and Education Association
61 Ethel Road West
Piscataway, New Jersey 08854

THE ESSENTIALS OF ADVANCED CALCULUS I ®

Printed in the United States of America

Library of Congress Catalog Card Number 87-61819

International Standard Book Number 0-87891-567-2

Revised Printing 1989

WHAT "THE ESSENTIALS" WILL DO FOR YOU

This book is a review and study guide. It is comprehensive and it is concise.

It helps in preparing for exams, in doing homework, and remains a handy reference source at all times.

It condenses the vast amount of detail characteristic of the subject matter and summarizes the **essentials** of the field.

It will thus save hours of study and preparation time.

The book provides quick access to the important facts, principles, theorems, concepts, and equations of the field.

Materials needed for exams, can be reviewed in summary form — eliminating the need to read and re-read many pages of textbook and class notes. The summaries will even tend to bring detail to mind that had been previously read or noted.

This "ESSENTIALS" book has been carefully prepared by educators and professionals and was subsequently reviewed by another group of editors to assure accuracy and maximum usefulness.

Dr. Max Fogiel
Program Director

CONTENTS

CONTENTS

CHAPTER 1

SETS AND THE NUMBER SYSTEMS

1.1 SETS

A set is a group or collection of objects. For example, the letters of the alphabet is the set: $S = \{A, B, C, D, \ldots, Z,\}$. The individual objects in the set are called members or elements.

1) Set Notation

It is a convention to denote sets with capital letters and members of sets with lower case letters.

a) $a \in A$: denotes that a is an element of the set A.

b) $a \notin A$: denotes that a is not an element of the set A.

c) ϕ : denotes the empty set, (the set without any elements)

d) to define the members of a set S we use the following expression

$S = \{X$: list the properties X must have to satisfy to belong to set $S\}$.

For example, the set of negative real numbers is

$$S = \{x : x \in \mathbb{R} : x < 0\}$$

e) $A \subset B$: denotes that every element of A is an element of B. For example, $\phi \subset A$.

2) Operations between two sets

Let A and B denote sets contained in U, the Universal Set. All operations are relative to this set U.

a) $A \cup B$: the set of all elements in either A or B (or both).

b) $A \cap B$: the set of all elements in both A and B.

c) $A - B$: the set of all elements in A and not in B.

d) $U - A$: the set of all elements not in A, commonly denoted $-A$.

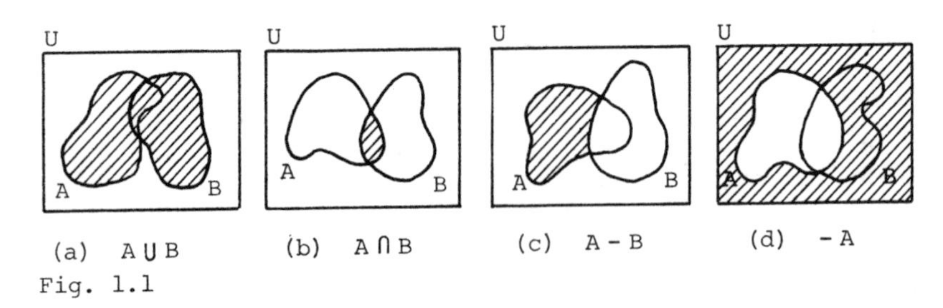

(a) $A \cup B$ (b) $A \cap B$ (c) $A - B$ (d) $-A$

Fig. 1.1

A pictorial explanation of these operations, known as a Venn diagram, is shown in Fig. 1.1.

1.2 THE FAMILIAR NUMBER SYSTEMS

a) $\mathbb{N}$ = the set of natural numbers, (1,2,3,..).

b) $\mathbb{Z}$ = the set of all integers, (...,-3,-2,-1,0,1,2,3,..).

c) $\mathbb{Q}$ = the set of all rational numbers, i.e.

$$\mathbb{Q} = \{x: x = \frac{a}{b}, a,b \in \mathbb{Z}, b \neq 0\}$$

d) $\mathbb{Q}'$ = the set of all irrational numbers. Irrational numbers such as as $\sqrt{2}$, and π are those real numbers which cannot be expressed as rational numbers

e) $\mathbb{R}$ = the set of real numbers. $\mathbb{R} = \mathbb{Q} \cup \mathbb{Q}'$.

1.3 PROPERTIES OF THE RATIONAL AND IRRATIONAL NUMBERS

The order and arithmetical properties of the rational and irrational numbers are listed in table 1.1.

Table 1.1 Properties of the real numbers

1) Order Properties

a) either $a > b$, $a = b$, or $a < b$ (order property of real numbers)

b) $a \leq b,\ b < c \Rightarrow a < c$ (transitive property of inequality)

$a < b,\ b \leq c \Rightarrow a < c$

2) Arithmetical properties

a) Addition $c = a + b$

$a + b = b + a$ (commutative law)

$(a + b) + c = a + (b + c)$ (associative law)

$a < b \Rightarrow a + c < b + c$ (addition property of inequality)

$a + 0 = a$ (effect of additive identity)

$a + (-a) = 0$ (effect of additive inverse)

b) Subtraction $c = b - a = b + (-a)$

c) Multiplication $p = ab$

$ab = ba$ (commutative law)

$(ab)c = a(bc)$ (associative law)

$a(b + c) = ab + ac$ (distributive law)

$a > b,\ c > 0 \Rightarrow ac > bc$

$a > b,\ c < 0 \Rightarrow ac < bc$

$a > b,\ c = 0 \Rightarrow ac = bc$

$a \cdot 1 = a$ (multiplicative identity)

$a \cdot a^{-1} = 1$ (multiplicative inverse)

d) Division $q = \frac{a}{b} = ab^{-1}$

3) Completeness

Between any two numbers there lies an infinite number of other real numbers. (see 1.6).

1.4 ABSOLUTE VALUE OF REAL NUMBERS

The absolute value of a real number a, denoted by $|a|$ is defined by

$$|a| = \begin{cases} a, & \text{if } a \geq 0 \\ -a, & \text{if } a < 0 \end{cases} \tag{1.1}$$

The main properties of $|a|$ are:

$$|a| = |-a|$$

$$\left|\frac{1}{a}\right| = \frac{1}{|a|}, \quad \text{if } a \neq 0$$

$$|ab| = |a|\,|b|, \text{ in general } |ab\ldots n| = |a|\,|b|\ldots|n|$$

$$\left|\frac{a}{b}\right| = \frac{|a|}{|b|}, \quad b \neq 0$$

$$\big||a| - |b|\big| \leq |a + b| \leq |a| + |b|$$

$$|a + b + \ldots + n| \leq |a| + |b| + \ldots + |n|$$

(triangular inequality)

1.5 GEOMETRY AND THE NUMBER SYSTEM

Some useful geometrical terms used in number systems are:

1) Point set: any collection of points.

2) One-dimensional point set: a set of points (real numbers) located on the real line.

3) Interval: a point set is sometimes described by inequalities of one of the following types:

   a) $a \leq x \leq b$, closed interval $[a,b]$

   b) $a \leq x < b$, half open interval $[a,b)$

   c) $a < x \leq b$, half closed interval $(a,b]$

   d) $a < x < b$, open interval (a,b) where a and b are called the end points of the interval. The length of the interval is $|b - a|$.

1.6 BOUNDED SETS

A non-empty set S, is defined as:

a) Bounded above, if there exists a number M such that $x \leq M$ for all $x \in S$.

b) Bounded below, if there exists a number m such that $x \geq m$ for all $x \in S$.

c) Bounded, if for all $x \in S$, $m \leq x \leq M$. If a set S is bounded above by M_0 and if when we find any other upper bound M, we find that $M \geq M_0$, then we say that M_0 is the least upper bound, supremum, for S. It is denoted by $M_0 = \sup_{x \in S} S$. Similarily, if the set S is bounded below by m_0, and if, whenever we find any other lower bound m, infimum, we find that $m \leq m_0$, then we say that m_0 is the greatest lower bound for S. It is denoted $m_0 = \inf_{x \in S} S$.

If the set S is a non-empty set of real numbers which has an upper bound, then S has a least upper bound. Similarly, if S has a lower bound, then it has a greatest lower bound. The rational numbers, Q, do not have these properties.

1.7 NEIGHBORHOODS

1) One-dimensional point set

A neighborhood of a point x_0 is the set of all points x lying within some fixed distance h, of x_0. That is, the points x such that $|x - x_0| < h$ or $x_0 - h < x < x_0 + h$, $h > 0$. A deleted neighborhood of a point x_0 is a neighborhood from which the point x_0 itself has been removed.

2) Two-dimensional point set

A rectangular neighborhood of the point (x_0, y_0) is the set of all points such that $|x - x_0| < h$ and $|y - y_0| < h$, $h > 0$. The set $|x - x_0| < h$, $|y - y_0| < h$ which excludes (x_0, y_0) is called a rectangular deleted h neighborhood of (x_0, y_0). A circular neighborhood of the point (x_0, y_0) is the set of all points (x,y) lying inside some circle with center at (x_0, y_0). That is, $\{(x,y) : \sqrt{(x-x_0)^2 + (y-y_0)^2} < h, h > 0\} = N_h(x_0, y_0)$.

1.8 OPEN AND CLOSED SETS

A two-dimensional set S is said to be *open* if each point of S has some neighborhood which belongs entirely to the set S.

The complement of a set S is the set of all points which do not belong to S. A set is *closed* if its complement is open. A point x is a boundary point of a set S if every neighborhood of x contains points in S and points in the complement of S. The set of boundary points of a set S is called its boundary, (denoted ∂S).

Theorems

a) The boundary of a set S and the boundary of its complement are the same.

b) If S is open then S contains none of its boundary points.

c) If S is closed, it contains all of its boundary points.

1.9 REGIONS

A two-dimensional region R is an open set any two of whose points can be connected by a line lying entirely in R. A closed region is a region containing all its boundary points.

1.10 THE SPACE R^n

If U is a given set, then $U \times U$, denoted by U^2, is the set of all ordered pairs of elements (u_1,u_2), $u_1,u_2 \in U$. By extension, U^n is the set of all ordered n-tuples of elements of

U, i.e. $U = \{(u_1, u_2, \ldots, u_n): u_i \in U, i = 1, 2, \ldots, n\}$. The set of real n-tuples, $R^n = \{(x_1, x_2, \ldots, x_n): x_i \in R, i = 1, 2, \ldots, n\}$, is called n-dimensional Euclidean space.

The set of all points whose distance from a center c is less than a radius $r > 0$ is called the ball in R^n centered at c of radius r, denoted by $B(c,r) = \{x \in R^n: d(x,c) < r\}$, where $d(x,c)$ denotes the distance between points x, and c.

The two-dimensional circular neighborhoods mentioned in (1.7) are balls in R^2.

1.11 COUNTABILITY

A set S is countable if it can be put into a one-to-one correspondence with a subset of the natural numbers, N. A set S is infinite if it can be placed into a one-to-one correspondence with a proper subset of itself. Proper means the subset is contained in, but not equal to the set.

An infinite set may or may not have a limit point. A finite set cannot have a limit point.

1.12 COMPLEX NUMBER SYSTEM

The algebraic system in which $x^2 + 1 = 0$ has a solution, namely, $x = \sqrt{-1}$, is the complex number system. Here we set $\sqrt{-1}$ equal to the symbol i.

1) Forms of a complex number

The expression $z = x + iy$ is called a complex number. The complex number z has the following forms:

a) Cartesian form

$z = x + iy$, where

$x, y \in R$

$i^2 = -1.$

b) Polar form

$z = \rho(\cos\theta + i\sin\theta)$, where

$x = \rho\cos\theta$

$y = \rho\sin\theta$

$\rho = |z| = \text{modulus of } z = \sqrt{x^2 + y^2}$

$$\theta = \arg z = \begin{cases} \tan^{-1} y/x; & x \neq 0 \\ \pi/2 + k\pi; & k \in \mathbb{Z},\ x = 0,\ y \neq 0 \\ \text{undefined}; & z = 0 \end{cases}$$

c) Euler's Formula

$e^{i\theta} = \cos\theta + i\sin\theta$; so $z = \rho e^{i\theta}$

d) Matrix form

The set of complex numbers is the collection of all 2 × 2 matrices of the form $\begin{pmatrix} a & -b \\ b & a \end{pmatrix}$, where

$$\begin{pmatrix} a & -b \\ b & a \end{pmatrix} = a\begin{pmatrix} 1 & 0 \\ 0 & 1 \end{pmatrix} + b\begin{pmatrix} 0 & -1 \\ 1 & 0 \end{pmatrix}, \text{ and } i = \begin{pmatrix} 0 & -1 \\ 1 & 0 \end{pmatrix}$$

2) Properties of complex numbers

If $z_1 = x_1 + iy_1 = \rho_1(\cos\theta_1 + i\sin\theta_1)$ and

$z_2 = x_2 + iy_2 = \rho_2(\cos\theta_2 + i\sin\theta_2)$ then:

a) $z_1 + z_2 = (x_1 + x_2) + i(y_1 + y_2)$

b) $z_1 - z_2 = (x_1 - x_2) + i(y_1 - y_2)$

c) $z_1 \cdot z_2 = (x_1x_2 - y_1y_2) + i(x_1y_2 + x_2y_1)$

$= \rho_1\rho_2\{\cos(\theta_1 + \theta_2) + i\sin(\theta_1 + \theta_2)\}$

d) $z_1/z_2 = \dfrac{(x_1x_2 + y_1y_2) + i(x_2y_1 - x_1y_2)}{x_2^2 + y_2^2}$

e) $z^n = \{\rho\,(\cos\theta + i\sin\theta)\}^n$

$= \rho^n(\cos(n\theta) + i\,\sin(n\theta))$

f) $|z_1 + z_2| \leq |z_1| + |z_2|$

g) $|z_1 - z_2| \geq \big||z_1| - |z_2|\big|$

3) Roots of a complex number

For each positive integer n, the complex number z has exactly n distinct roots given by the relation

$$z^{\frac{1}{n}} = \rho^{\frac{1}{n}}\left\{\cos\left(\frac{\theta + 2k\pi}{n}\right) + i\sin\left(\frac{\theta + 2k\pi}{n}\right)\right\}$$

where $k = 0,1,2,\ldots,n - 1$

4) Complex Conjugate

The number $\bar{z} = a - ib$ is called the conjugate of the complex number $z = a + ib$. The following rules apply to complex conjugates:

a) $\overline{z_1 + z_2} = \bar{z}_1 + \bar{z}_2$

b) $\overline{z_1 - z_2} = \bar{z}_1 - \bar{z}_2$

c) $\overline{z_1 z_2} = \overline{z_1}\;\overline{z_2}$

d) $\overline{\left(\dfrac{z_1}{z_2}\right)} = \dfrac{\overline{z_1}}{\overline{z_2}}$

5) Point set theory in the complex plane

A circular neighborhood of a point z_o is an open interval containing z_o, equivalently $N(z_o,\varepsilon) = \{z: |z - z_o| < \varepsilon\}$.

1.13 MATHEMATICAL INDUCTION

The principle of mathematical induction is useful in proving propositions involving all positive integers. The method of proving a proposition, p(n), consists of the following steps:

a) Prove that p(n) is true for n = 1.

b) Assume p(n) true for n = k, where k is any positive integer.

c) From the assumption in (b) prove that p(n) must be true for n = k+ 1.

d) Then p(n) is true for every natural number n.

CHAPTER 2

ELEMENTARY FUNCTIONS AND SEQUENCES

2.1 FUNCTIONS

A function consists of two sets X and Y and a rule which establishes a correspondence between the two sets.

Namely, a function is a set of ordered pairs (x,y) with $x \in X$ and $y \in Y$. If to each value of the variable x corresponds one value of a variable y, we call y a function of x (or, $f: X \to Y$), and we donate the function by $y = f(x)$. The set X is called the domain of f and the set Y is called the range of f (Fig. 2.1).

Since the value of y is determined by x, it is called the dependent variable; and since x is free to take on any value in X, it is called the independent variable.

The graph of a function is a two-dimensional picture of a point set given by (x,y) or $(x,f(x))$, where x is in the domain and y or $f(x)$ is the corresponding value. If only one value of y corresponds to each value of x in the domain, the function is called single valued. If more than one value of y corresponds to some value of x, the function is called multiple valued.

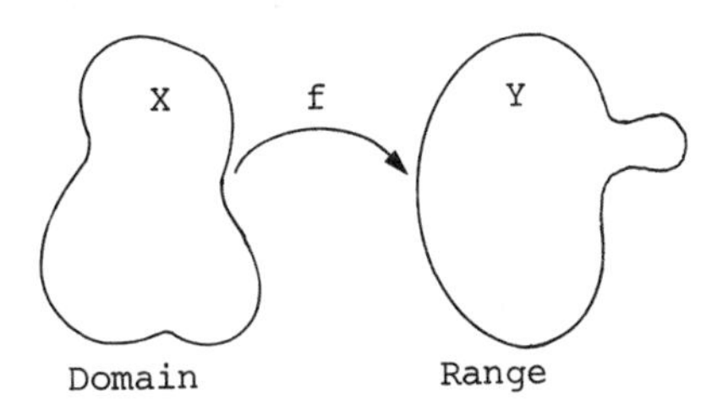

Fig. 2.1

2.2 BOUNDED FUNCTIONS

A real valued function is called bounded above if $f(x) \leq M$ for all x in the domain of f. The constant M is called an upper bound of the function.

If $f(x) \geq m$ for all x in the domain, the function $f: X \rightarrow Y$ is bounded below. The constant m is called a lower bound of the fucntion. If $m \leq f(x) \leq M$ for all x in the domain of f, we call f(x) bounded.

2.3 MONOTONIC FUNCTIONS

A function $f: X \rightarrow Y$ is called:

a) weakly monotonically increasing if for all x_1, x_2 such that $x_1 < x_2 \Rightarrow f(x_1) < f(x_2)$.

b) strictly monotonically increasing if for all x_1, x_2 such that $x. < x_2 \Rightarrow f(x_1) < f(x_2)$.

c) weakly monotonically decreasing if for all x_1, x_2 such that $x_1 < x_2 \Rightarrow f(x_1 > f(x_2)$.

d) strictly monotonically decreasing if for all x_1, x_2 such that $x_1 < x_2 \Rightarrow f(x_1) > f(x_2)$.

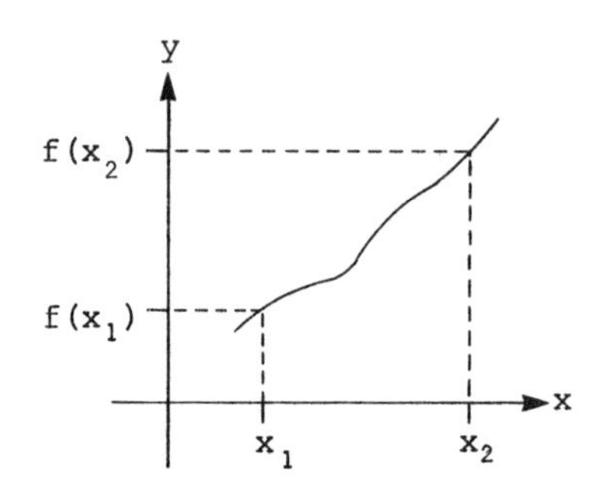

a. Monotonically increasing

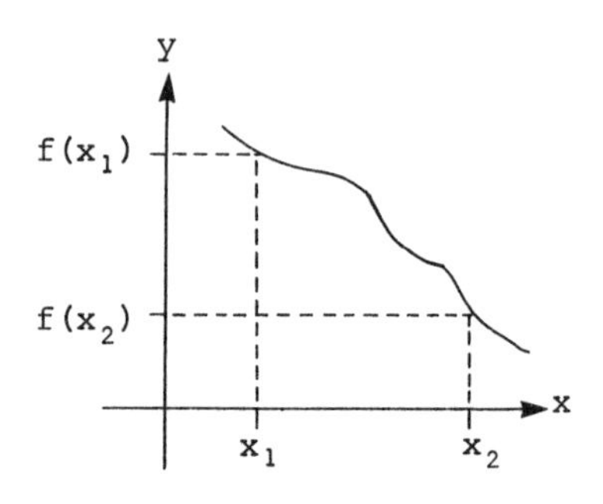

b. Monotonically decreasing

Fig. 2.2

2.4 INVERSE FUNCTION

If $f:X \to Y$, such that for each $y \in Y$, there is one and only one $x \in X$ for which $f(x) = y$, then define $g:Y \to X$ to be $g(y) = x$ if and only if $f(x) = y$. This is called the inverse function of f, and is denoted $g = f^{-1}$.

Theorem:

If $f:X \to Y$ is a strictly increasing continuous function in a closed interval, $a \leq x \leq b$, with $f(a) = a$, and $f(b) = b$ then there is a unique inverse function $g:Y \to X$ on $a = f(a) \leq y \leq f(b) = b$, which is strictly increasing and continuous.

2.5 MAXIMA AND MINIMA

Let $f:X \to Y$ be a function which is defined on an open interval (a,b), and let x_{01} be a point such that $a < x_{01} < b$, and x_{02} be a point such that $a < a_1 < x_{02} < b_1 < b$. Then the function has:

a) an absolute maximum if, for all x in X, $f(x) \leq f(x_{01})$.

b) a relative maximum if, for all x such that $a_1 < x < b_2$, $f(x) \leq f(x_{02})$.

c) an absolute minimum if, for all x in X, $f(x) \geq f(x_{01})$.

d) a relative minimum if, for all x such that $a_1 < x < b_1$ $f(x) \geq f(x_{02})$.

Theorem:

If a function $f: X \to Y$ has a relative minimum or maximum at a point x_o of the open interval (a,b), then if f is differentiable at x_o, $f'(x_o) = 0$.

2.6 LIMITS OF FUNCTIONS

Let f(x) be a real valued function defined for all values of x near x_o with the possible exception of x_o itself, then the limit of the function f(x) is defined as $\lim_{x \to x_0} f(x) = L$,

if and only if, for all $\varepsilon > 0$, there exists $\delta^o > 0$, such that, $|f(x) - L| < \varepsilon$ whenever $0 < |x - x_o| < \delta$.

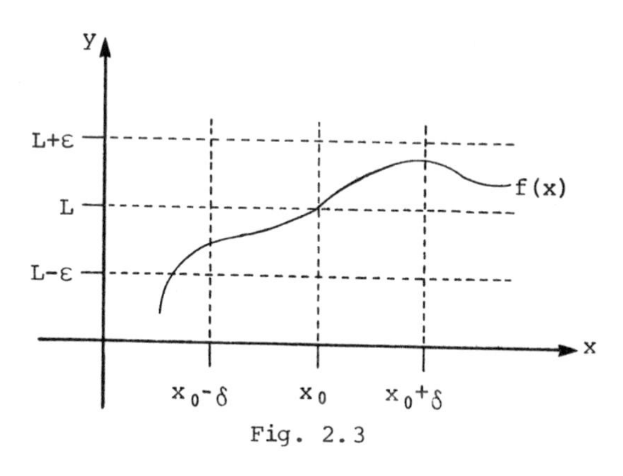

Fig. 2.3

1) Right-hand Limit

If x approaches x_o from the right $(x \to x_o^+)$, then, $f(x_o^+) = \lim_{x \to x_o^+} f(x) = \lim_{\substack{h>0 \\ h\to 0}} f(x_o + h) = L_1$

2) Left-hand Limit

If x approaches x_o from the left ($x \to x_o^-$), then, $f(x_o^-) = \lim_{x \to x_o^-} f(x) = \lim_{\substack{h>0 \\ h\to 0}} f(x_o - h) = L_2$

3) a) We have $\lim_{x\to x_o} f(x) = L$ if and only if $\lim_{x\to x_o^+} f(x) = \lim_{x \to x_o^-} f(x)$,

or $L_1 = L_2$.

b) If $f(x) \leq g(x) \leq h(x)$ for all x and $\lim_{x\to x_o} f(x) = \lim_{x\to x_o} h(x) = L \Rightarrow \lim_{x\to x_o} g(x) = L$ (Squeeze law).

c) $\lim_{x\to x_o} |f(x)| = \left| \lim_{x\to x_o} f(x) \right|$

d) if $f(x) > 0$, then $\lim_{x\to x_o} f(x) \geq 0$

e) if $f(x) > 0$, then $\lim_{x\to x_o} \sqrt{f(x)} = \sqrt{\lim_{x\to x_o} f(x)}$

f) if $\lim_{x\to x_o} f(x) = L_1$ and $\lim_{x\to x_o} g(x) = L_2$, then

1) $\lim_{x\to x_o} [f(x) \pm g(x)] = \lim_{x\to x_o} f(x) + \lim_{x\to x_o} g(x) = L_1 + L_2$

2) $\lim_{x\to x_o} [f(x) \cdot g(x)] = \lim_{x\to x_o} f(x) \cdot \lim_{x\to x_o} g(x)$

$= L_1 L_2$

3) $\lim_{x\to x_o} \frac{f(x)}{g(x)} = \frac{\lim_{x\to x_o} f(x)}{\lim_{x\to x_o} g(x)} = \frac{L_1}{L_2}$, if $L_2 \neq 0$

4) if $f(x) \geq g(x)$, then $\lim_{x\to x_o} f(x) \geq \lim_{x\to x_o} g(x)$

5) $\lim_{x \to x_0} \frac{1}{f(x)} = \frac{1}{L_1}, \quad \lim_{x \to x_0} \frac{1}{g(x)} = \frac{1}{L_2}, \quad$ if $L_1 \neq 0$, $L_2 \neq 0$.

4) Additional Definitions

a) $\lim_{x \to x_0} f(x) = +\infty$, if for all $M > 0$, there exists $\delta > 0$, such that $f(x) > M$ whenever $0 < |x - x_0| < \delta$

b) $\lim_{x \to x_0} f(x) = -\infty$ if for all $M > 0$ there exists $\delta > 0$, such that $f(x) < -M$ whenever $0 < |x - x_0| < \delta$

c) $\lim_{x \to +\infty} f(x) = L$ if for all $\varepsilon > 0$ there exists $N > 0$, such that $|f(x) - L| < \varepsilon$ whenever $x > N$.

A similar formulation can be made for $\lim_{x \to -\infty} f(x)$

5) Special Limits

a) $\lim_{x \to 0} \frac{\sin x}{x} = 1$

b) $\lim_{x \to 0} \frac{1 - \cos x}{x} = 0$

c) $\lim_{x \to +\infty} \left(1 + \frac{1}{x}\right)^x = e$

d) $\lim_{x \to 0^+} (1 + x)^{\frac{1}{x}} = e$

e) $\lim_{x \to 0} \frac{e^x - 1}{x} = 1$

f) $\lim_{x \to 1} \frac{x - 1}{\ln x} = 1$

2.7 CONTINUITY OF FUNCTIONS

1) Right- and left-hand continuity. The right-hand limit of a function f(x) at x_o is defined as

$$f(x_o^+) = \lim_{x \to x_o^+} f(x) = \lim_{h \to 0^+} f(x+h) = L \qquad (2.1)$$

The left-hand limit is defined as

$$f(x_o^-) = \lim_{x \to x_o^-} f(x) = \lim_{\substack{h \to 0 \\ h<0}} f(x+h) = \lim_{h \to 0^+} f(x-h) \qquad (2.2)$$

2) Continuous functions

A function f(x), defined for all values of x near $x = x_o$ as well as at $x = x_o$, is called continuous at x_o if both left- and right-hand limits exist, are equal and

$$\lim_{x \to x_o} f(x) = f(x_o) \qquad (2.3)$$

i.e. the function is continuous at x_o if and only if for all $\varepsilon > 0$ there exists $\delta > 0$ such that $|f(x) - f(x_o)| < \varepsilon$ whenever $|x - x_o| < \delta$, and $f(x_o)$ equals the limit value. If for each δ there is an ε which holds for all points x_o of the interval, then f(x) is uniformly continuous on the interval. Points where f(x) fails to be continuous are called discontinuities of f(x), and f(x) is said to be discontinuous at these points. See figures 2.4 and 2.5.

3) Continuity in an interval if f(x) is defined in a closed interval $a \le x \le b$, then f(x) is continuous in the interval if and only if

a) $\lim_{x \to x_o} f(x) = f(x_0)$ for all x in (a,b)

b) $\lim_{x \to a^+} f(x) = f(a)$

c) $\lim_{x \to b^-} f(x) = f(b)$

A function f in a closed interval is called sectionally continuous or piecewise continuous (see Fig. 2.6) if the interval can be split up into finite number of subintervals such that in each subinterval the two following conditions hold.

- f(x) is continuous in each subinterval
- f(x) possesses (finite) limits at the left-hand and right-hand ends of each subinterval, i.e. $f(x_i^+)$ and $f(x_i^-)$ exist for all $i = 1,2,3,\ldots,n - 1$ and also, $f(a^+)$ and $f(b^-)$ exist.

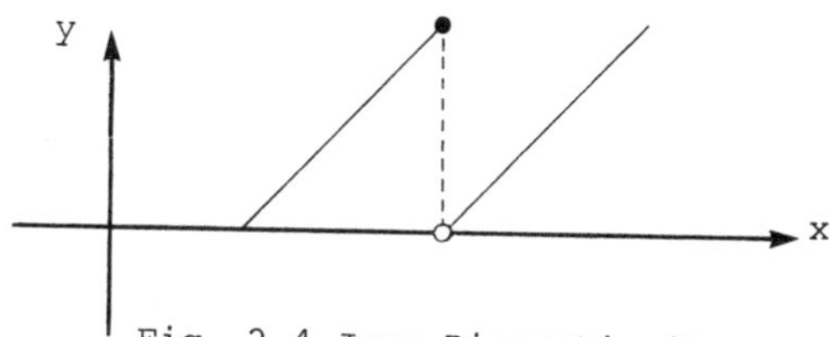

Fig. 2.4 Jump Discontinuity

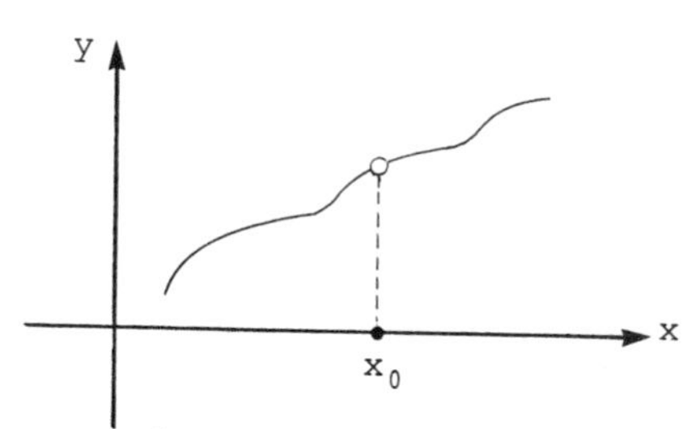

Fig. 2.5 Removable discontinuity

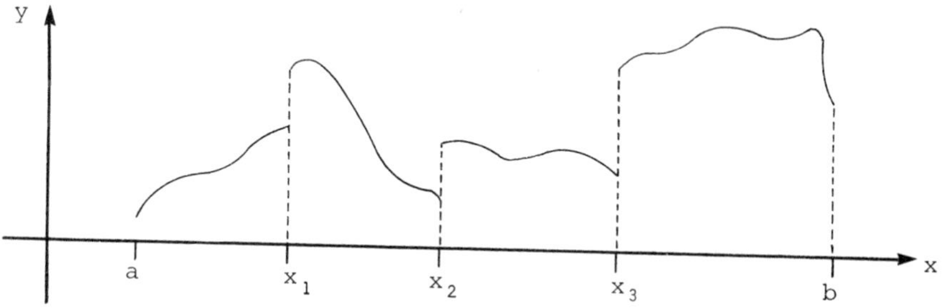

Fig. 2.6 Sectionally Continuous

4) Theorems

a) if $f(x)$ and $g(x)$ are two continuous functions at $x = x_o$ then

1) $f(x) + g(x)$, $f(x) - g(x)$, and $f(x)g(x)$ are continuous at x_o.

2) if $g(x_o) \neq 0$ $\frac{f(x)}{g(x)}$ is continuous at x_o.

3) if $f(x) > 0$, $\sqrt{f(x)}$ is continuous at x_o.

4) $|f(x)|$ is continuous at x_o.

5) the composite function $g[f(x)]$ is continuous at x_o

b) if $f(x)$ is continuous on $[a,b]$, it is bounded on that interval.

c) if $\lim_{x \to x_o} f(x) = f(x_o)$ on $[a,b]$, i.e. $f(x)$ is continuous on $[a,b]$, $f(a) = A$, $f(b) = B$, and $A \neq B$, then for each c, $(A < c < B)$, there exists a point x_o in $[a,b]$ for which $f(x_o) = c$ (Intermediate Value Theorem).

d) if $f(x)$ is also strictly monotonic, then there is exactly one x_o for which $f(x_o) = c$.

e) if $A > 0$, $B < 0$ or $A < 0$, $B > 0$, then there is a point x_o in $[a,b]$ for which $f(x_o) = 0$.

f) if $\lim_{x \to x_o} f(x) = f(x_o)$, f continuous on $[a,b]$, then there exists an x in $[a,b]$ such that $f(x) = M$ or $f(x) = m$, where M is the least upper bound, and m is the greatest lower bound.

g) The following functions are continuous for all $x \in R$:

1) polynomials: $f(x) = a_o + a_1 x + \ldots + a_n x^n$

2) $\sin x$ and $\cos x$

3) a^x, $a > 0$

4) Any sum, product, composition, etc. of the above. This is a result of theorems in 2.7.4.

2.8 SEQUENCES

1) Definition of a sequence

A sequence, S_n, is a particular kind of function, $f:Z^+ \to Y$, whose independent variable n ranges over the set of positive integers. Thus a seuqence, $\{S_n\}$, is a set of real numbers $s_1, s_2, s_3 \ldots$ in a one-to-one correspondence with the natural numbers.

2) Limits of a sequence

The limit L of a sequence S_n is defined as: $\lim_{n \to \infty} S_n = L$, if and only if, for all $\varepsilon > 0$, there exists an N, such that $|S_n - L| < \varepsilon$ for all $n > N$.

If the limit of the sequence exists, the sequence is called convergent; otherwise it is called divergent.

$\lim_{n \to \infty} S = \infty$ if for all $M > 0$, there exists an $N > 0$, such that $S_n > M$, for all $n > N$.

$\lim_{n \to \infty} S_n = -\infty$ if for all $M > 0$, there exists an $N > 0$ such that $S_n < -M$, for all $n > N$.

3) Bounded Sequences

A sequences, S_n, is called bounded above if $S_n \leq M$, for $n = 1,2,3,\ldots.$ The constant M (independent of n) is called an upper bound. If $S_n \geq m$ for all $n = 1,2,3,\ldots$ the sequence is bounded below, and m is called the lower bound. If $m \leq S_n \leq M$ the sequence is called bounded.

4) Monotonic Sequences

A sequence, $\{S_n\}$, is called:

a) monotonically increasing, if $S_{n+1} \geq S_n$ for all $n \in N$

b) strictly monotonically increasing, if $S_{N+1} > S_N$ for al $n \in N$.

c) monotonically decreasing if $S_{N+1} \leq S_N$ for all $n \in N$

d) strictly monotonically decreasing if $S_{n+1} < S_n$ for all $n \in N$

Theorem: Every bounded monotonic sequence of real numbers has a limit.

CHAPTER 3

DERIVATIVES AND INTEGRALS

3.1 DEFINITION OF THE DERIVATIVE

Let f(x) be a real valued function defined at any point x_0: $a < x_0 < b$. The derivative of f at x is defined as

$$\boxed{\lim_{h\to 0} \frac{f(x_0+h) - f(x_0)}{h}} \tag{3.1}$$

and is denoted $f'(x_0)$ or $\frac{df}{dx}(x_0)$.

The function f(x) is called differentiable at the point $x = x_0$ if it has a derivative at this point.

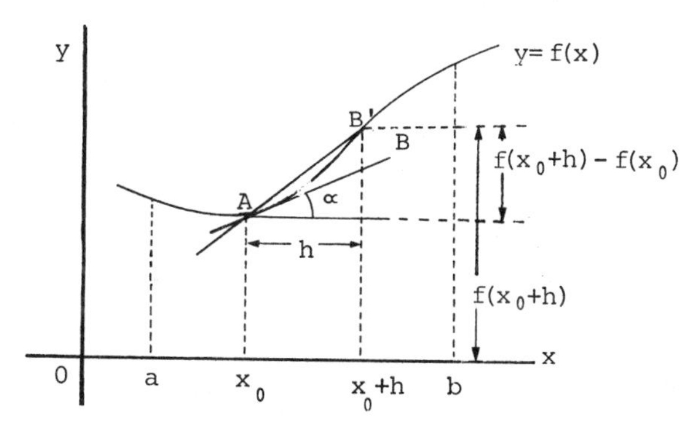

Fig. 3.1 Graphical Interpretation of the derivative.

Graphically the derivative represents the slope of the tangent line AB to the function at the point $(x_0, f(x_0))$, or

$$\tan\alpha = \lim_{h \to 0} \frac{f(x_0 + h) - f(x_0)}{h} \tag{3.2}$$

The equation for the tangent line to the curve $y = f(x)$ at the point x_0 is given by

$$y - f(x_0) = f'(x_0)(x - x_0) \tag{3.3}$$

If $y - f(x_0) = dy$, and $x - x_0 = dx$, then the above equation becomes $dy = f'(x_0)dx$. We call dx and dy differentials; dy is the differential of f(x) at x_0. A function is infinitely differentiable, when it has derivatives of all orders $(f'(x), f''(x), \ldots)$. The derivative of a function, f(x), can be interpreted as the rate of change of y with respect to x.

1) Right-Hand Derivative

   The right-hand derivative of f(x) at $x = x_0$ is defined as

$$f'_+(x_0) = \lim_{h \to 0^+} \frac{f(x_0 + h) - f(x_0)}{h}, \tag{3.4}$$

2) Left-Hand Derivative

   The left-hand derivative of f(x) at $x = x_0$ is defined as

$$f'_-(x_0) = \lim_{h \to 0^-} \frac{f(x_0 + h) - f(x_0)}{h}, \tag{3.5}$$

3.2 RULES OF DIFFERENTIATION

If f(x) and g(x) are differentiable functions, the following rules are valid:

If f(x), g(x) and h(x) are differentiable functions, the following rules are valid:

$$\frac{d}{dx}\{f(x) + g(x)\} = f'(x) + g'(x)$$

$$\frac{d}{dx}\{f(x) - g(x)\} = f'(x) - g'(x)$$

$$\frac{d}{dx}\{cf(x)\} = cf'(x), \quad c \text{ is any constant}$$

$$\frac{d}{dx}\{f(x)g(x)\} = f(x)g'(x) + f'(x)g(x)$$

$$\frac{d}{dx}\frac{f(x)}{g(x)} = \frac{f'(x)g(x) - f(x)g'(x)}{[g(x)]^2}, \qquad g(x) \neq 0$$

If $y = f[g(x)]$, then $\frac{dy}{dx} = \frac{dy}{du} \cdot \frac{du}{dx} = f'[g(x)]g'(x)$

If $y = f(x)$, and there is an inverse function $(x = f^{-1}(y))$, then

$$\frac{dy}{dx} = \frac{1}{dx/dy}$$

If $x = f(t)$, $y = g(t)$, then

$$\frac{dy}{dx} = \frac{dy/dt}{dx/dt} = \frac{g'(t)}{f'(t)}$$

3.3 THEOREMS OF DIFFERENTIABLE FUNCTIONS

A) If $f(x)$ is differentiable at x_0, it is continuous there.

B) If $f(x)$ is continuous on the closed interval $[a,b]$, then there is a point $x' \in [a,b]$ for which

$$f(x') < f(x) \quad (x: x \in [a,b])$$

C) If $f(x)$ is continuous on the closed interval $[a,b]$, then there is a point $x.$ in $[a,b]$ for which

$$f(x_0) \geq f(x) \quad (x: x \in [a,b])$$

D) If $f(x)$ is an increasing function on an interval, then at each point x_0, where $f(x)$ is differentiable we have

$$f'(x_0) \geq 0$$

E) If $f(x)$ is strictly increasing on an interval, and suppose also that $f'(x_0) > 0$ for some x_0 in the interval, then the inverse function $f^{-1}(x)$ if it exists, is differentiable at the point $y_0 = f(x_0)$.

F) If $f(x)$ is differentiable on the interval $[a,b]$, and $g(x)$ is a differntiable function in the range of f, then the composed function $h = g \circ f$ ($h(x) = g[f(x])$ is also differentiable on $[a,b]$.

G) Suppose that $f(x)$, $g(x)$ are differentiable on the closed interval $[a,b]$ and that $f'(x) = g'(x)$ for all $x \in [a,b]$, then there is a constant c such that $f(x) = g(x) + C$.

H) Rolle's Theorem

If $f(x)$ is continuous on $[a,b]$, differentiable on (a,b), and $f(a) = f(b) = 0$, then there is a point ζ in (a,b) such that $f'(\zeta) = 0$.

I) Mean Value Theorem

a) f is continuous on $[a,b]$

b) f is differentiable on (a,b) then there exists some point $\zeta \in (a,b)$, such that $f'(\zeta) = \dfrac{f(b) - f(a)}{b - a}$.

J) Cauchy's Theorem of the Mean

If $f(x)$ and $g(x)$ are continuous on $[a,b]$ and differentiable on (a,b), then there exists a ζ in (a,b) such that

$$\frac{f(b) - f(a)}{g(b) - g(a)} = \frac{f'(\zeta)}{g'(\zeta)}, \quad \text{where}$$

$g(a) \neq g(b)$ and $g'(\zeta) \neq 0$

K) Taylor's Theorem of the Mean

If $f(x)$ and its first $n + 1$ derivatives are continuous on $[a,b]$, then for any x and any c in (a,b), we have

$$f(x) = f(c) + f'(c)(x - c) + \ldots + \frac{f^{(n)}(c)(x - c)^n}{n!} + R_{n+1},$$

where

$$R_{n+1} = \frac{1}{n!} \int_c^x (x - t)^n f^{n+1}(t)dt.$$

The remainder, R_{n+1}, can be written in the following forms:

1)

$$R_{n+1} = \frac{f^{n+1}(d)(x-c)^{n+1}}{(n+1)!}, \quad \text{where}$$

$d \in (a,b)$, or

2)

$$R_{n+1} = \frac{f^{n+1}(d)(x-d)^{n}(x-c)}{n!}, \quad d \in (a,b).$$

L) Let $f(x)$ be differentiable on (a,b) and $f'(b) = 0 = f'(a)$ then for some point β on (a,b) we have:

1) If $f'(x) \geq 0,\ x < \beta$
$f'(x) \leq 0,\ x > \beta$ $\Longrightarrow f(\beta) \geq f(x),\ (a < x < b)$.

2) If $f'(x) > 0,\ x < \beta$
$f'(x) < 0,\ x > \beta$ $= f(\beta) > f(x),\ (x \neq \beta,\ a<x<b)$.

M) Let $f(x)$ have n continuous derivatives on (a,b) and suppose that $x = \beta$, we have

$$f(\beta) = 0,\quad f'(\beta) = 0 \ldots f^{n-1}(\beta) = 0, \quad \text{and} \quad f^{(n)}(\beta) \neq 0.$$

1) $f(x)$ has a maximum at $x = \beta$, only if $f^{(n)}(b) < 0$.

2) $f(x)$ has a minimum at $x = b$, only if $f^{(n)}(b) > 0$.

N) L'Hôpital's Rule

Let $f(x)$ and $g(x)$ be differentiable at the points in the interval (a,b) except possibly at a point x_0 in (a,b) and $g(x) \neq 0$ for $x = x_0$. If $f(x) \to 0$ and $g(x) \to 0$ as $x \to x_0$ or if $|f(x)| \to \infty$ and $|g(x)| \to \infty$ as $x \to x_0$, then

$$\boxed{\lim_{x \to x_0} \frac{f(x)}{g(x)} = \lim_{x \to x_0} \frac{f'(x)}{g'(x)}}$$

, if this limit exists.

This equation can be extended to cases where $x \to \infty$ or $-\infty$ or where $x_0 = a$ or b, or in which only one-sided limits such as $x_0 \to a^+$, $x_0 \to b^-$ are involved.

3.4 DEFINITE INTEGRALS

Let f(x) be a bounded function on [a,b] and P an increasing sequence of points $a < x_1 < x_2 < \ldots < x_n < b$.

Let $\Sigma = \sum_{K=1}^{n} M_K(x_K - x_{K-1})$ and $\sigma = \sum_{k=1}^{m} m_k(x_k - x_{k-1})$ be the sums corresponding to the approximations to the area under the graph of f(x) given in Figure 3.2, where M_K, m_k are the maximum and minimum values of f(x) on the interval $[x_{k-1}, x_k]$.

If $\inf_P \Sigma = \sup_P \sigma$, the function f(x) is Riemann integrable.

The common value is called the definite integral of f(x) over the interval [a, b] and is denoted by $\int_a^b f(x)dx$.

In this symbol, f(x)dx is often called the integrand, and $a \leq x \leq b$ is called the range of integration.

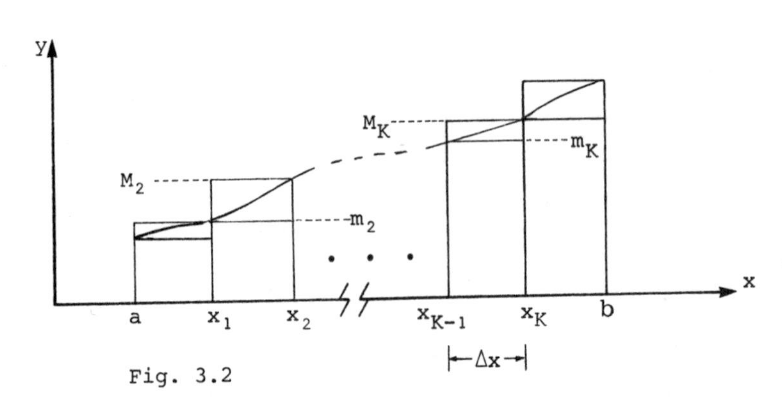

Fig. 3.2

Theorems of Definite Integrals

If f(x) and g(x) are integrable on [a,b], then:

A) $$\int_a^b \{f(x) \pm g(x)\}dx = \int_a^b f(x)dx \pm \int_a^b g(x)dx.$$

B) $$\int_a^b cf(x)dx = c\int_a^b f(x)dx, \quad c = \text{any constant.}$$

C) $$\int_a^b f(x)dx = \int_a^c f(x)dx + \int_c^b f(x)dx, \quad a < c < b.$$

D) $$\int_a^b f(x)dx = -\int_b^a f(x)dx.$$

E) $$\int_a^a f(x)dx = 0 = \int_b^b f(x)dx.$$

F) If $x \in [a,b]$ and $f(x) \in [c_1,c_2]$ for all x in $[a,b]$, then

$$c_1(b - a) \le \int_a^b f(x)dx \le c_2(b - a).$$

G) If $x \in [a,b]$, and $f(x) \le g(x) \implies$

$$\int_a^b f(x)dx \le \int_a^b g(x)dx.$$

H) $$\left| \int_a^b f(x)dx \right| \le \int_a^b |f(x)|dx.$$

I) If $f(x)$ and $g(x)$ are integrable on $[a,b]$, then so is their product $f(x) \cdot g(x)$.

J) First Mean Value Theorem

If $f(x)$ is continuous on $[a,b]$, there is a point ζ on (a,b) such that

$$\frac{1}{b - a}\int_a^b f(x)dx = f(\zeta).$$

K) If f(x) and g(x) are continuous on $a \leq x \leq b$, and g(x) has a constant sign (+ or -) in the interval, then there is a point ζ such that

$$\int_a^b f(x)g(x)dx = f(\zeta) \int_a^b g(x)dx.$$

L) Bonnet's Theorem

If f(x) and g(x) are continuous on [a,b] and if g(x) is monotonically increasing or decreasing, there is a point ζ on (a,b) such that

$$\int_a^b f(x)g(x)dx = g(a) \int_a^{\zeta} f(x)dx + g(b) \int_{\zeta}^b f(x)dx.$$

M) Fundamental Theorem of Calculus

If f(x) is continuous on the interval [a,b] such that $F'(x) = f(x)$, then the integral $\int_a^b f(x)$ exists for all $x \in [a,b]$ and

$$\boxed{\int_a^b f(x)dx = F(b) - F(a).}$$

N) Schwarz's Inequality

If f(x) and g(x) are continuous,

$$\left(\int_a^b f(x)g(x)dx \right)^2 \leq \int_a^b \{f(x)\}^2 dx \int_a^b \{g(x)\}^2 dx.$$

3.5 INDEFINITE INTEGRALS

Let f(x) be a real valued function on the closed interval [a,b]. If F(x) is the differentiable function defined on [a,b] such that $F(x) = 0$ and $F'(x) = f(x)$ for all $x \in [a,b]$, then $\int f(x)$ or $F(x)$ is called the indefinite intergral of $f(x)$.

An indefinite integral can be expressed as a definite integral with a variable upper limit, or

$$F(x) = \int f(x)dx = \int_a^x f(t)dt + c, \text{ or} \tag{3.6}$$

$$F'(x) = \frac{d}{dx}\left[\int_a^x f(t)dt + c\right] = f(x),$$

with t instead of x in some places because putting x as both a variable of integration and a limit of integration is ambiguous. t is called a dummy variable.

3.6 METHODS OF INTEGRATION

A) Integration by Parts

If $f(x)$ and $g(x)$ are continuous and have continuous derivatives on $[a,b]$, then

$$\int f(x)g'(x)dx = f(x)g(x) - \int f'(x)g(x)dx$$

B) Partial Fractions

Any rational function, $f(x) = \frac{P(x)}{Q(x)}$, where $P(x)$ and $Q(x)$ are polynomials, can be written as the sum of rational functions having the form $\frac{A}{(ax+b)^r}$, $\frac{Ax+B}{(ax^2+bx+c)^r}$ where $r = 1,2,3,\ldots n$.

C) Change of Variable

Integrals like $\int f(x)dx$ can be written as $\int f\{g(t)\}g'(t)dt$, using the transformation $x = g(t)$.

3.7 NUMERICAL METHODS FOR EVALUATING DEFINITE INTEGRALS

A) Rectangular Rule

$$\int_a^b f(x)dx = \Delta x \{y_1 + y_2 + y_3 + \ldots + y_n\} \tag{3.7}$$

where $y_i = f(a + i\,\Delta x)$

B) Trapezoidal Rule

$$\int_a^b f(x)dx = \frac{\Delta x}{2} \{y_0 + 2y_1 + 2y_2 + \ldots + 2y_{n-1} + y_n\} \tag{3.8}$$

where $y_i = f(a + i\Delta x)$

C) Simpson's Rule

$$\int_a^b f(x)dx = \frac{\Delta x}{3} \{y_0 + 4y_1 + 2y_2 + 4y_3 + 2y_4 + \ldots + 2y_{2n-2} + 4y_{2n-1} + y_{2n}\} \tag{3.9}$$

where $y_i = f(a + i\,\Delta x)$

D) Taylor's Theorem of the Mean

See theorem 3.3.K, Derivatives, page 26

3.8 STIELTJES INTEGRAL

Let $P = \{x_0, x_1, x_2, \ldots, x_n\}$ be a partition of $[a,b]$, and let $x_1', x_2', \ldots, x_n'$ be points in the respective subintervals (x_0, x_1), $(x_1, x_2), \ldots, (x_{n-1}, x_n)$. Then the Stieltjes integral is defined as

$$\int_a^b f(x)dg(x) = \lim_{|P| \to 0} \sum_{i=1}^{n} f(x'_i)[g(x_i) - g(x_{i-1})] \qquad (3.10)$$

where $|P|$ denotes the mesh fineness of the partition P, i.e. $\max_i |x_{i+1} - x_i|$.

3.9 ELLIPTIC INTEGRALS

The integrals

$$F(k, \phi) = \int_0^\phi \frac{d\theta}{\sqrt{1 - k^2 \sin\theta}}, \quad E(k, \phi) = \int_0^\phi \sqrt{1 - k^2 \sin^2\theta}\, d\theta$$

and (3.11)

$$M(k, n, \phi) = \int_0^\phi \frac{d\theta}{(1 + n\sin^2\theta)\sqrt{1 - k^2 \sin\theta}},$$

where $0 < k < 1$, are known as elliptic integrals of the first, second and third kind respectively. ϕ is called the amplitude, written $\phi = \mathrm{am}(u)$, and k is its modulus, written $k = \mathrm{mod}(u)$. These integrals have differentiation rules and simplifcation rules similar to sine and cosine.

CHAPTER 4

FUNCTIONS OF SEVERAL VARIABLES, PARTIAL DERIVATIVES

4.1 FUNCTIONS OF SEVERAL VARIABLES

A more general case of a function may be illustrated by considering a function of, say, three variables, x, y, and z

$$f = f(x,y,z).$$

If $z = f(x,y)$, then z is a dependent variable and x, y are the independent variables.

The function $z = f(x,y)$ is single-valued if only one value of z corresponds to each pair (x,y) for which the function is defined. The same function is called multiple-valued, if there is more than one value of z corresponding to each pair (x,y), A function which is multiple-valued can be considered as a collection of single-valued functions.

4.2 LIMITS AND CONTINUITY

Let f(x,y) be a function defined in a region R such that (x_0,y_0) is either an interior point of R or on the boundary of R, then the limit, L, of the function $f(x,y)$ is defined as

$\lim_{(x,y) \to (x_0,y_0)} f(x,y) = L$, if and only if, for all $\varepsilon > 0$, there exists $\delta > 0$ such that

$$|f(x,y) - L| < \varepsilon \text{ whenever } 0 < |x - x_0| < \delta, \text{ and}$$

$$0 < |y - y_0| < \delta.$$

The theorems on limits for functions of two or more variables are the same as the theorems of functions of one variable.

The function $f(x,y)$ is continuous at (x_0,y_0) if

$$\lim_{(x,y) \to (x_0,y_0)} \to f(x,y) \to f(x_0,y_0) \text{ i.e.}$$

$\lim_{(x,y) \to (x_0,y_0)} f(x,y) = f(x_0,y_0)$ if and only if for all $\varepsilon > 0$ there exists a $\delta > 0$ such that

$$|f(x,y) - f(x_0,y_0)| < \varepsilon \quad \text{if} \quad |x - x_0| < \delta \quad \text{and}$$

$$|y - y_0| < \delta.$$

The function $f(x,y)$ is uniformly continuous in R if the δ's depend only on the ε's, but not on any particular point (x_0,y_0) in R.

4.3 PARTIAL DERIVATIVES

If the real-valued function, $f:\mathbb{R}^n \to \mathbb{R}$, is defined in a neighborhood of $x_0 = (x_{01}, x_{02}, \ldots, x_{0n})$, then the n partial derivatives of $f:\mathbb{R}^n \to \mathbb{R}$ at x_0 are defined by

$$\frac{\partial f}{\partial x_i} = \left. \frac{df(x_{01}, x_{02}, \ldots, x_{0i-1}, x_{0i}, x_{0i+1}, \ldots, x_{0n})}{dx_i} \right|_{x_i = x_{0i}} \tag{4.1}$$

$$= \lim_{t \to 0} \frac{f(x_{01}, x_{02}, \ldots, x_{0i}+t, \ldots, x_{0n}) - f(x_{01}, x_{02}, \ldots, x_{0n})}{t}$$

$$(i = 1,2,3,\ldots,n).$$

Thus, the partial derivative is computed simply by considering all but the relevant variable as constant.

The partial derivatives of f(x,y) with respect to x and y are denoted by

$$\frac{\partial f}{\partial x} \quad \left(\text{or } f_x,\ f_x(x,y),\ \left.\frac{\partial f}{\partial x}\right|_y\right)$$

and

$$\frac{\partial f}{\partial y} \quad \left(\text{or } f_y,\ f_y(x,y),\ \left.\frac{\partial f}{\partial y}\right|_x\right)$$

respectively.

By definition

$$\frac{\partial f}{\partial x} = \lim_{\Delta x \to 0} \frac{f(x + \Delta x, y) - f(x,y)}{\Delta x}, \text{ and} \qquad (4.2)$$

$$\frac{\partial f}{\partial y} = \lim_{\Delta y \to 0} \frac{f(x, y + \Delta y) - f(x,y)}{\Delta y}. \qquad (4.3)$$

The partial derivatives of f(x,y) at a point (x_0, y_0) are

$$\frac{\partial f}{\partial x}(x_0,y_0) = \lim_{h \to 0} \frac{f(x_0 + h,\ y_0) - f(x_0,y_0)}{h} \qquad (4.4)$$

and

$$\frac{\partial f}{\partial y}(x_0,y_0) = \lim_{h \to 0} \frac{f(x_0,y_0 + h) - f(x_0,y_0)}{h} \qquad (4.5)$$

provided these limits exist.

If f(x,y) has partial derivatives at each point (x,y) in a region, then $\frac{\partial f}{\partial x}$ and $\frac{\partial f}{\partial y}$ may themselves have partial derivatives. These second derivatives are denoted by

$$\frac{\partial}{\partial x}\left(\frac{\partial f}{\partial x}\right) = \frac{\partial^2 f}{\partial x^2} = f_{xx},\quad \frac{\partial}{\partial y}\left(\frac{\partial f}{\partial y}\right) = \frac{\partial^2 f}{\partial y^2} = f_{yy} \ldots \text{ etc.}$$

In a similar manner, higher order derivatives may be defined. For example,

$$\frac{\partial}{\partial y}\left(\frac{\partial^2 f}{\partial x^2}\right) = \frac{\partial^3 f}{\partial y\, \partial x^2} = f_{xxy}$$

4.4 DIFFERENTIALS

Let $f(x,y)$ be a real-valued function of the variables x and y.

If $\Delta x = dx$ and $\Delta y = dy$ are increments of x and y respectively, then

$$\Delta f = \Delta z = f(x + \Delta x,\ y + \Delta y) - f(x,y) \tag{4.6}$$

is the increment in $z = f(x,y)$.

The total differential of $f(x,y)$ is defined by the equation

$$df = \frac{\partial f}{\partial x} dx + \frac{\partial f}{\partial y} dy \tag{4.7}$$

In the case of three variables, x, y, z the total differential of $f(x,y)$ may be defined by the equation

$$df = \frac{\partial f}{\partial x} dx + \frac{\partial f}{\partial y} dy + \frac{\partial f}{\partial z} dz . \tag{4.8}$$

If $f:\mathbb{R}^n \to \mathbb{R}$, then

$$df = \frac{\partial f}{\partial x_1} dx_1 + \frac{\partial f}{\partial x_2} dx_2 + \ldots + \frac{\partial f}{\partial x_n} dx_n, \tag{4.9}$$

where $f:\mathbb{R}^n \to \mathbb{R}$ is defined in a neighborhood of $x = (x_1,x_2,\ldots,x_n)$.

If $f(x_1,x_2,\ldots,x_n) = c$, a constant, then $df = 0$.

Exact Differentials

The expression $Pdx + Qdy$ is an exact differential of the function $f(x,y)$ if and only if $\frac{\partial P}{\partial y} = \frac{\partial Q}{\partial x}$. (4.10)

Also, the expression $Pdx + Qdy + Rdz$ is an exact differential of the function $f(x,y,z)$ if and only if

$$\frac{\partial P}{\partial y} = \frac{\partial Q}{\partial x},\quad \frac{\partial Q}{\partial z} = \frac{\partial R}{\partial y},\quad \frac{\partial R}{\partial x} = \frac{\partial P}{\partial z} \tag{4.11}$$

Here P, Q, R denote functions of x, y and z. If $Pdx + Qdy = 0$ where this is an exact differential, then $P\hat{i} + Q\hat{j}$ is a vector field which is the gradient of some real-valued function.

4.5 COMPOSITE FUNCTIONS AND THE CHAIN RULE

Let $u = F(x,y,z)$ be a given function where $x = f(t)$, $y = g(t)$, $z = h(t)$.

The composite function, $G(t)$, is defined by substituting these functions x,y,z in the function $u = F(x,y,z)$, or

$$G(t) = F(f(t),g(t),h(t)) \tag{4.12}$$

The formula of differentiation for this composite function is

$$\frac{du}{dt} = \frac{\partial u}{\partial x}\frac{dx}{dt} + \frac{\partial u}{\partial y}\frac{dy}{dt} + \frac{\partial u}{\partial z}\frac{dz}{dt} \tag{4.13}$$

It is apparent that $\frac{dx}{dt}$, $\frac{dy}{dt}$, and $\frac{dz}{dt}$ are just $f'(t)$, $g'(t)$, and $h'(t)$, respectively. If the variables x, y, and z are functions of two variables, s and t, then

$$G(s,t) = F[x(s,t),y(s,t),z(s,t)] \quad \text{and}$$

$$\frac{\partial G}{\partial t} = \frac{\partial u}{\partial x}\frac{\partial x}{\partial t} + \frac{\partial u}{\partial y}\frac{\partial y}{\partial t} + \frac{\partial u}{\partial z}\frac{\partial z}{\partial t} \tag{4.14}$$

In general, if $u = F(x_1,x_2,x_3,\ldots,x_n)$ where $x_1 = f_1(t_1,t_2,t_3,\ldots,t_p)$, ..., $x_n = f_n(t_1,t_2,t_3,\ldots,t_p)$ then

$$\frac{\partial u}{\partial t_k} = \frac{\partial u}{\partial x_1}\frac{\partial x_1}{\partial t_k} + \frac{\partial u}{\partial x_2}\frac{\partial x_2}{\partial t_k} + \ldots + \frac{\partial u}{\partial x_n}\frac{\partial x_n}{\partial t_k} \tag{4.15}$$

$$k = 1,2,\ldots,p$$

These results are called chain rules.

4.6 HOMOGENEOUS FUNCTIONS, EULER'S THEOREM

A function $f:\mathbb{R}^n \to \mathbb{R}$, $f(x_1,x_2,x_3,\ldots,x_n)$, is homogeneous of degree n if

$$f(tx_1,tx_2,tx_3,\ldots,tx_n) = t^n f(x_1,x_2,x_3,\ldots,x_n)$$

for all values of $x_1,x_2,x_3,\ldots,x_n$ and t for which $f(x_1,x_2,x_3,\ldots,x_n)$ and $f(tx_1,tx_2,tx_3,\ldots,tx_n)$ are defined.

Euler's Theorem

If $f(x_1,x_2,x_3,\ldots,x_n)$ is positively homogeneous of degree n, then

$$x_1 \frac{\partial f}{\partial x_1} + x_2 \frac{\partial f}{\partial x_2} + x_3 \frac{\partial f}{\partial x_3} + \ldots + x_n \frac{\partial f}{\partial x_n} = nf(x_1,x_2,x_3,\ldots,x_n), \tag{4.16}$$

where positively homogeneous means homogeneous for $t > 0$ only.

4.7 IMPLICIT FUNCTIONS

An equation of the form $F(x_1,x_2,\ldots,x_n) = 0$ defines one variable, say x_n, as a function of the remaining variables, $x_n = f(x_1,x_2,\ldots,x_{n-1})$ in some region about any point where $F(x_1,x_2,\ldots,x_n) = 0$ is satisfied and where $\frac{\partial f}{\partial x_n} \neq 0$. Then x_n is called an implicit function of $x_1,x_2,\ldots,x_{n-1}$.

The implicit function, f, can sometimes be solved for without knowing F, if enough of the partial derivatives of F are known. To do this, we use the chain rule.

4.8 JACOBIAN DETERMINANTS

Suppose that x_1 and x_2 are related by two equations of the form

$$F_1(x_1,x_2) = 0$$
$$F_2(x_1,x_2) = 0. \tag{4.17}$$

The Jacobian determinant of F_1 and F_2 with respect to x_1 and x_2 is

$$\frac{\partial (F_1,F_2)}{\partial (x_1,x_2)} = \begin{vmatrix} \frac{\partial F_1}{\partial x_1} & \frac{\partial F_1}{\partial x_2} \\ \frac{\partial F_2}{\partial x_1} & \frac{\partial F_2}{\partial x_2} \end{vmatrix}. \tag{4.18}$$

This is called a Jacobian of second order. The form of a general nth order Jacobian involves n functions, each of n variables:

$$\frac{\partial(F_1,F_2,\ldots,F_n)}{\partial(x_1,x_2,\ldots,x_n)} = \begin{vmatrix} \frac{\partial F_1}{\partial x_1} & \frac{\partial F_1}{\partial x_2} & \cdots & \frac{\partial F_n}{\partial x_n} \\ \frac{\partial F_2}{\partial x_1} & \cdots & \cdots & \cdots \\ \vdots & \vdots & \vdots & \vdots \\ \frac{\partial F_n}{\partial x_1} & \cdots & \cdots & \frac{\partial F_n}{\partial x_n} \end{vmatrix}. \tag{4.19}$$

The Jacobians can be used in obtaining partial derivatives of implicit functions. Thus given the simultaneous equations $F_1(x_1,x_2,x_3,x_4) = 0$, $F_2(x_1,x_2,x_3,x_4) = 0$, we may consider $x_1 = f(x_3,x_4)$ and $x_2 = f(x_3,x_4)$, then

$$\frac{\partial x_1}{\partial x_3} = \frac{-\frac{\partial(F_1,F_2)}{\partial(x_3,x_2)}}{\frac{\partial(F_1,F_2)}{\partial(x_1,x_2)}} \qquad \frac{\partial x_1}{\partial x_4} = \frac{\frac{\partial(F_1,F_2)}{\partial(x_4,x_2)}}{\frac{\partial(F_1,F_2)}{\partial(x_1,x_2)}}$$

$$\frac{\partial x_2}{\partial x_3} = \frac{-\frac{\partial(F_1,F_2)}{\partial(x_1,x_3)}}{\frac{\partial(F_1,F_2)}{\partial(x_1,x_2)}} \qquad \frac{\partial x_2}{\partial x_4} = \frac{-\frac{\partial(F_1,F_2)}{\partial(x_1,x_4)}}{\frac{\partial(F_1,F_2)}{\partial(x_1,x_2)}}. \tag{4.20}$$

These ideas can be easily extended to simultaneous equations of n variables. Thus if

$F_1(x_1,x_2,\ldots,x_n) = 0, \ldots, F_n(x_1,x_2,\ldots,x_n) = 0$ and if

$x_1 = f(x_{n-1},x_n)$, $x_2 = f(x_{n-1},x_n)$, ..., $x_{n-2} = f(x_{n-1},x_n)$, then

$$\frac{\partial x_1}{\partial x_n} = -\frac{\dfrac{\partial(F_1,F_2,\ldots,F_n)}{\partial(x_n,x_2,x_3,\ldots,x_{n-2})}}{\dfrac{\partial(F_1,F_2,\ldots,F_n)}{\partial(x_1,x_2,\ldots,x_{n-2})}}, \ldots, \frac{\partial x_{n-2}}{\partial x_n} = -\frac{\dfrac{\partial(F_1,F_2,\ldots,F_n)}{\partial(x_n,x_1,x_2,\ldots,x_{n-3})}}{\dfrac{\partial(F_1,F_2,\ldots,F_n)}{\partial(x_1,x_2,\ldots,x_n)}}$$

(4.21)

$$\frac{\partial x_1}{\partial x_{n-1}} = -\frac{\dfrac{\partial(F_1,F_2,\ldots,F_n)}{\partial(x_{n-1},x_2,x_3,\ldots,x_{n-2})}}{\dfrac{\partial(F_1,F_2,\ldots,F_n)}{\partial(x_1,x_2,\ldots,x_{n-2})}}, \ldots, \frac{\partial x_{n-2}}{\partial x_{n-1}} = -\frac{\dfrac{\partial(F_1,F_2,\ldots,F_n)}{\partial(x_{n-1},x_1,x_2,\ldots,x_{n-3})}}{\dfrac{\partial(F_1,F_2,\ldots,F_n)}{\partial(x_1,x_2,\ldots,x_{n-2})}}.$$

4.9 TRANSFORMATIONS OF VARIABLES

A mapping or transformation of variables is a function which establishes a correspondence between two sets of variables.

If u, v are the variables of one set, and x, y are the variables of a second set, then the equations

$$x = f(u,v)$$

$$y = g(u,v) \qquad (4.21)$$

where f and g are functions, define a transformation between points in the uv and the xy plane.

Solving (4.42) for u and v the transformation $u = f^*(x,y)$, $v = g^*(x,y)$ is obtained, which is called the inverse transformation corresponding to (4.42).

The Jacobians $\frac{\partial(u,v)}{\partial(x,y)} = \frac{\partial(f^*,g^*)}{\partial(x,y)}$ and $\frac{\partial(x,y)}{\partial(u,v)} = \frac{\partial(f,g)}{\partial(u,v)}$ of these transformations are reciprocals of each other. The above ideas can be extended to transformations in three or higher dimensions.

These Jacobians play an important role in the change of variables in multiple integrals.

4.10 TAYLOR SERIES

Taylor's Theorem of the Mean

A function of two or more variables often can be expanded in a power series which generalizes the familiar one-dimensional expansion.

If all the nth partial derivatives of f(x,y) are continuous in a closed region and if the (n + 1) partial derivatives exist in the open region, then

$$f(x+h,\ y+k) = \sum_{k=0}^{n-1} \frac{1}{k!}\left(h\frac{\partial}{\partial x} + k\frac{\partial}{\partial y}\right)^k f(x,y) + R_n, \tag{4.22}$$

where R_n, the remainder after n terms is given by

$$R_n = \frac{1}{n!}\left(h\frac{\partial}{\partial x} + k\frac{\partial}{\partial y}\right)^n f(x + \tau h, y + \tau k) \tag{4.23}$$

$(0 < \tau < 1)$.

Series (4.43) is called a Taylor series in two variables. Extensions to three or more variables can be made.

4.11 MAXIMA AND MINIMA

If a point $P(x_0, y_0)$ is an interior point of a region in which all partial second derivatives exist and are continuous, then at that point f has:

A) a relative maximum if

$$\frac{\partial f}{\partial x} = \frac{\partial f}{\partial y} = 0, \quad \frac{\partial^2 f}{\partial x^2} < 0 \text{ and } \left(\frac{\partial^2 f}{\partial x^2}\right)\left(\frac{\partial^2 f}{\partial y^2}\right) > \left(\frac{\partial^2 f}{\partial x \partial y}\right)^2$$

at P.

B) a relative minimum if

$$\frac{\partial f}{\partial x} = \frac{\partial f}{\partial y} = 0, \quad \frac{\partial^2 f}{\partial x^2} > 0 \text{ and } \left(\frac{\partial^2 f}{\partial x^2}\right)\left(\frac{\partial^2 f}{\partial y^2}\right) > \left(\frac{\partial^2 f}{\partial x \partial y}\right)^2 \text{ at P.}$$

C) a saddle point (neither a relative maximum nor a relative minimum) if

$$\left(\frac{\partial^2 f}{\partial x^2}\right)\left(\frac{\partial^2 f}{\partial y^2}\right) < \left(\frac{\partial^2 f}{\partial x\, \partial y}\right)^2$$

When $\left(\frac{\partial^2 f}{\partial x^2}\right)\left(\frac{\partial^2 f}{\partial y^2}\right) = \left(\frac{\partial^2 f}{\partial x\, \partial y}\right)^2$ at P,

further investigation is necessary.

4.12 CONSTRAINTS AND LAGRANGE MULTIPLIERS

A method to find the relative maxima and minima of a real-valued function of n variables subject to m constraints,

$$g_1(x_1,\ldots,x_n) = 0,\ g_2(x_1,\ldots,x_n) = 0,\ldots,g_m(x_1,\ldots,x_n) = 0,$$

is by noticing that at x, a relative maxima or minima, we must have

$$\frac{\partial f}{\partial x_1}(x) = \lambda_1 \frac{\partial g_1}{\partial x_1}(x) + \lambda_2 \frac{\partial g_2}{\partial x_1}(x) + \ldots + \lambda_m \frac{\partial g_m}{\partial x_1}(x)$$

$$\frac{\partial f}{\partial x_2}(x) = \lambda_1 \frac{\partial g_1}{\partial x_2}(x) + \lambda_2 \frac{\partial g_2}{\partial x_2}(x) + \ldots + \lambda_m \frac{\partial g_m}{\partial x_2}(x)$$

$$\vdots$$

$$\frac{\partial f}{\partial x_n}(x) = \lambda_1 \frac{\partial g_1}{\partial x_n}(x) + \lambda_2 \frac{\partial g_2}{\partial x_n}(x) + \ldots + \lambda_m \frac{\partial g_m}{\partial x_n}(x).$$

In the language of vector analysis, we have

$$\nabla f = \lambda_1 \nabla g_1 + \lambda_2 \nabla g_2 + \ldots + \lambda_m \nabla g_m.$$

E.g. If $n = 3$, $m = 1$, we are finding the maxima and minima of a three-dimensional scalar field constrained to a surface. Then

$$\nabla f = \lambda \nabla g.$$

Note: Any maxima and minima which exist under the constraints must satisfy the above properties, but not all points satisfying these properties are maxima or minima.

4.13 DIFFERENTIATION OF INTEGRALS INVOLVING A PARAMETER (LEIBNITZ'S RULE)

If $\phi(x)$ is defined by the integral:

$$\phi(x) = \int_a^b f(x,t)dt,$$

in some interval $c < x < d$ such that in the rectangle $\{x \in (c,d),\ t \in (a,b)\}$ $f_x(x,t)$ exists and is continuous.

Then

$$\boxed{\phi'(x) = \int_a^b \frac{\partial}{\partial x}(f(x,t))dt.}$$

In other words, you can pass the derivative inside the integral.

This is called Leibnitz's Rule.

Using the Chain Rule, the Fundamental Theorem of Calculus, and Leibnitz's Rule, we get:

if

$$\phi(x) = \int_{A(x)}^{B(x)} f(x,t)dt,$$

then

$$\phi'(x) = \int_{A(x)}^{B(x)} \frac{\partial}{\partial x}(f(x,t))dt + f(x,B(x))B'(x) - f(x,A(x))A'(x),$$

assuming the same conditions as above and that A and B are differentiable.

CHAPTER 5

VECTOR ANALYSIS

5.1 ELEMENTARY PROPERTIES OF VECTORS

A) Vectors and Scalars

A scalar quantity is one which possesses only magnitude, whereas a vector quantity possesses both magnitude and direction. Thus the position of a point B (terminal point) relative to a point A (initial point) can be completely described by a vector $\vec{AB}$ from A to B. The magnitude or length of the vector is denoted by $|\vec{AB}|$. We denote vectors either by putting an arrow over them or by capital letters. A three-dimensional vector is an ordered triplet of real numbers (A_1, A_2, A_3). (Fig. 5.1)

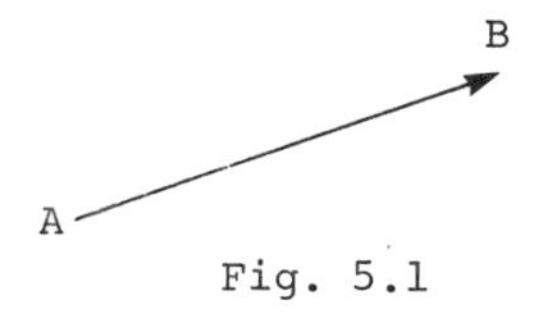

Fig. 5.1

B) Vector Algebra

a) The sum or resultant of vectors $\vec{A}$ and $\vec{B}$ is a vector $\vec{C}$ such that $\vec{C} = \vec{A} + \vec{B}$. (Fig. 5.2)

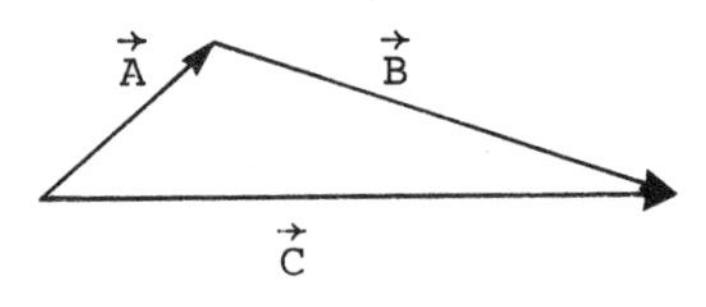

Fig. 5.2

b) The difference of vector $\vec{A}$ and $\vec{B}$, $\vec{A} - \vec{B}$, is that vector $\vec{C}$ which added to $\vec{B}$ gives $\vec{A}$. If $\vec{A} = \vec{B}$, then $\vec{A} - \vec{B}$ is defined as the null or zero vector. It has a zero magnitude but its direction is not defined.

c) The multiplication of a vector $\vec{A}$ by a scalar m produces a vector with magnitude $\vec{C} = m|\vec{A}|$ and direction the same or opposite to that of $\vec{A}$ according as m is positive or negative. If m = 0, $m\vec{A} = \vec{0}$

d) Two vectors are equal if they have the same direction and the same magnitude.

e) A vector of length one is called a unit vector. The rectangular unit vectors i, j, k are unit vectors having the direction of the positive x, and y, and z axes of a rectangular coordinate system. (Fig. 5.3)

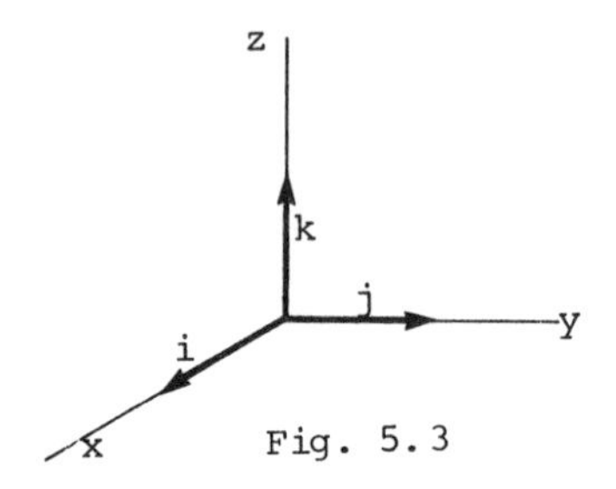

Fig. 5.3

f) Any vector $\vec{A}$ whose projections on the x, y, and z axes are a_x, a_y, and a_z, respectively, can be written as the vector sum

$$\vec{A} = a_x i + a_y j + a_z k$$

where a_x, a_y, and a_z are often called the coordinates of $\vec{A}$.

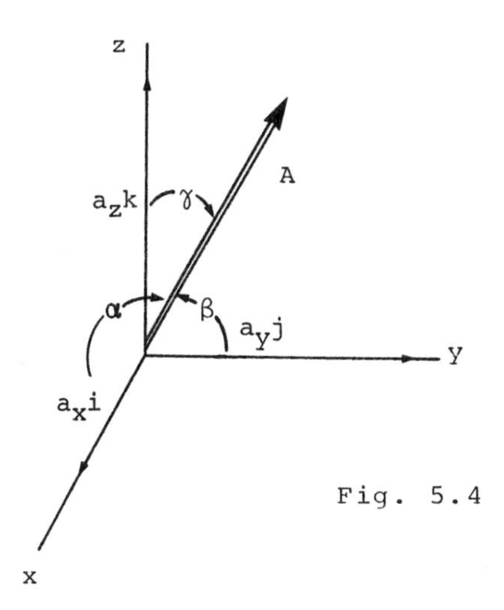

Fig. 5.4

If α, β, γ are the angles measured to A from the positive x, y and z axes respectively, then

$$a_x = |A| \cos \alpha$$

$$a_y = |A| \cos \beta$$

$$a_z = |A| \cos \gamma$$

The numbers $\cos\alpha$, $\cos\beta$, and $\cos\gamma$ are called the direction cosines of A and are frequently denoted by l, m and n. Any three numbers proportional to l, m, and n are known as 'direction ratios'.

From figure 5.4 the following equations can be easily derived:

1) $|\vec{A}| = \sqrt{a_x^2 + a_y^2 + a_z^2}$.

2) $l^2 + m^2 + n^2 = \cos^2\alpha + \cos^2\beta + \cos^2\gamma = 1$.

3) $\cos\alpha = \dfrac{a_x}{|\vec{A}|} = l$, $\quad \cos\beta = \dfrac{a_y}{|\vec{A}|} = m$,

$$\cos\gamma = \frac{a_z}{|\vec{A}|} = n.$$

g) The laws of vector algebra are listed in table 5.1, where $\vec{A}$, $\vec{B}$ and $\vec{C}$ are vectors and m, n are scalars.

Table 5.1 Laws of Vector Algebra	
Commutative Law	$\vec{A} + \vec{B} = \vec{B} + \vec{A}$
Associative Law	$\vec{A} + (\vec{B} + \vec{C}) = (\vec{A} + \vec{B}) + \vec{C}$
Associative Law	$m(n\vec{A}) = (mn)\vec{A} = n(m\vec{A})$
Distributive Law	$(m + n)\vec{A} = m\vec{A} + n\vec{A}$
Distributive Law	$m(\vec{A} + \vec{B}) = m\vec{A} + m\vec{B}$

5.2 THE SCALAR OR DOT PRODUCT OF TWO VECTORS

The dot or scalar product of two vectors $\vec{A}$ and $\vec{B}$, written as $\vec{A} \cdot \vec{B}$, is defined to be a scalar quantity equal to the product of the lengths of the two vectors and the cosine of the angle θ between the positive directions of the vectors.

It is given by the equation

$$\vec{A} \cdot \vec{B} = |A||B| \cos\theta, \tag{5.1}$$

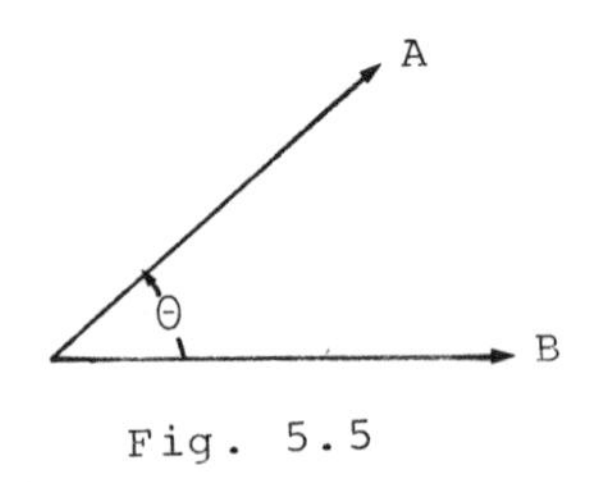

Fig. 5.5

The following laws are valid:

$\vec{A} \cdot \vec{B} = \vec{B} \cdot \vec{A}.$

$\vec{A} \cdot (\vec{B} + \vec{C}) = \vec{A} \cdot \vec{B} + \vec{A} \cdot \vec{C}.$

$i \cdot j = j \cdot i = j \cdot k = k \cdot j = i \cdot k = k \cdot i = 0,\ i \cdot i = j \cdot j = k \cdot k = 1.$

If $\vec{A} = a_x i + a_y j + a_z k$, $\vec{B} = b_x i + b_y j + b_z k$, then

$$\vec{A} \cdot \vec{B} = a_x b_x + a_y b_y + a_z b_z,$$

$$\vec{A} \cdot \vec{A} = a_x^2 + a_y^2 + a_z^2 = |A|^2,\ \vec{B} \cdot \vec{B} = b_x^2 + b_y^2 + b_z^2 = |\vec{B}|^2,$$

$$\cos\theta = \frac{\vec{A} \cdot \vec{B}}{|\vec{A}|\,|\vec{B}|} = \frac{a_x b_x + a_y b_y + a_z b_z}{\sqrt{a_x^2 + a_y^2 + a_z^2}\sqrt{b_x^2 + b_y^2 + b_z^2}},$$

$$= l_1 l_2 + m_1 m_2 + n_1 n_2.$$

If $\vec{A} \cdot \vec{B} = 0$, then $\vec{A}$ and $\vec{B}$ are perpendicular.

5.3 THE VECTOR OR CROSS PRODUCT OF TWO VECTORS

The vector or cross product of two vectors $\vec{A}$ and $\vec{B}$, written as $\vec{A} \times \vec{B}$ or [AB], is defined to be a vector $\vec{C}$ with magnitude equal to the product of the lengths of the two vectors and the sine of the angle θ between them, and direction perpendicular to the plane of $\vec{A}$ and $\vec{B}$. It is given by the equation

$$|\vec{C}| = |\vec{A} \times \vec{B}| = |\vec{A}||\vec{B}|\sin\theta \quad (0 \leq \theta \leq n). \tag{5.2}$$

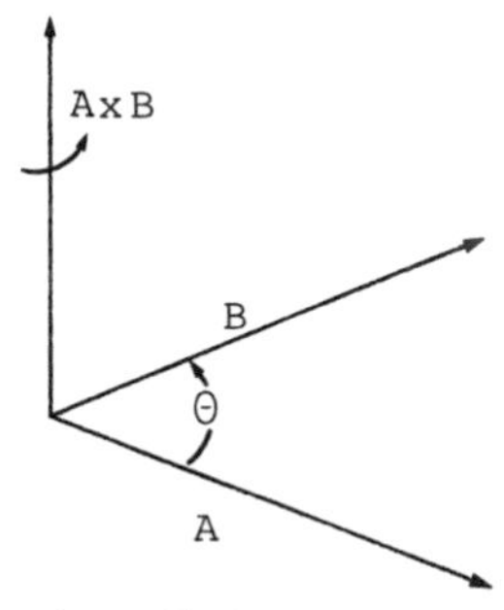

Fig. 5.6

The vector A × B is so oriented that A is rotated into B about A × B by the right-hand rule.

The following laws are valid:

$\vec{A} \times \vec{B} = -\vec{B} \times \vec{A}$.

$\vec{A} \times (\vec{B} + \vec{C}) = \vec{A} \times \vec{B} + \vec{A} \times \vec{C}$.

$i \times j = k, \quad j \times k = i, \quad k \times i = j, \quad i \times i = j \times j = k \times k = 0.$

If $\vec{A} = a_x i + a_y j + a_z k$, $\vec{B} = b_x i + b_y j + b_z k$, then

$$\vec{A} \times \vec{B} = \det \begin{vmatrix} i & j & k \\ a_x & a_y & a_z \\ b_x & b_y & b_z \end{vmatrix}$$

$$= (a_y b_z - a_z b_y)i + (a_z b_x - a_x b_z)j + (a_x b_y - a_y b_x)k.$$

If $\vec{A} \times \vec{B} = 0$, then $\vec{A}$ and $\vec{B}$ are parallel, or at least one of $\vec{A}$ or $\vec{B}$ has length zero.

5.4 MULTIPLE PRODUCTS

Three types of products involving three vectors (A,B,C) are of importance. These products are:

A) The scalar triple product $(A \times B) \cdot C$

B) The vector triple product $(A \times B) \times C$ and

C) The $(A \cdot B)C$ triple product

The following laws are valid:

$(A \cdot B)C \neq A(B \cdot C)$;

$(A \times B) \cdot C = (B \times C)A = (C \times A)B$;

$(A \times B) \times C \neq A \times (B \times C)$;

$A \times (B \times C) = (A \cdot C)B - (A \cdot B)C;$

$(A \times B) \times C = (A \cdot C)B - (B \cdot C)A;$

If $A = a_x i + a_y j + a_z k$, $B = b_x i + b_y j + b_z k$ and

$C = c_x i + c_y j + c_z k$, then

$$A \cdot (B \times C) = \det \begin{vmatrix} a_x & a_y & a_z \\ b_x & b_y & b_z \\ c_x & c_y & c_z \end{vmatrix}.$$

5.5 VECTOR SPACES

A collection V of vectors is called a real vector space if there is a sum $\vec{v} + \vec{w} \in V$ defined for all members $\vec{v}$ and $\vec{w}$ of V and also a product $r\vec{v} \in V$ of a real number r and any vector $\vec{V}$ such that the following properties hold:

$\vec{v} + \vec{w} = \vec{w} + \vec{v}$ (commutative law)

$\vec{u} + (\vec{v} + \vec{w}) = (\vec{u} + \vec{v}) + \vec{w}$ (associative law)

$\vec{v} + \vec{0} = \vec{v}$

$s(\vec{v} + \vec{w}) = s\vec{v} + s\vec{w}$

$s(t\vec{v}) = (st)\vec{v}$

$1\vec{v} = \vec{v}$

$0\vec{v} = \vec{0}.$

5.6 VECTOR FUNCTIONS

If a vector, $\vec{V}$, is associated to a point t in some

domain D, then $\vec{V}$ is called a vector valued function of t in D. It is denoted by F and is given symbolically by

$$\vec{V} = F(t). \tag{5.3}$$

If the domain, D, is a region in two, three, or more dimensions, the vector function is often called a vector field.

In three dimensions we can write,

$$\vec{V}(t) = V_1(t)i + V_2(t)j + V_3(t)k, \tag{5.4}$$

and for a three-dimensional domain, to each point (x,y,z) in the xyz-system corresponds a vector indicated by,

$$\vec{V}(x,y,z) = V_1(x,y,z)i + V_2(x,y,z)j + V_3(x,y,z)k. \tag{5.5}$$

5.7 CONTINUITY AND DERIVATIVES OF VECTOR FUNCTIONS

A) Continuity

Let f(x) be a function defined for all values of x near $t = t_0$ as well as at $t = t_0$. Then the function f(x) is said to be continuous at t_0 if

$$\lim_{t \to t_0} f(t) = f(t_0) \tag{5.6}$$

or, equivalently,

$\lim_{t \to t_0} f(t) = f(t_0)$ if and only if for all $\varepsilon > 0$, there exists a $\delta > 0$, such that $|f(t) - f(t_0)| < \varepsilon$, if $|t - t_0| < \delta$.

B) Derivative

The derivative of the vector valued function V(t) with respect to $t \in \mathbb{R}$ is defined as the limit

$$\frac{dV(t)}{dt} = \lim_{\Delta t \to 0} \frac{V(t + \Delta t) - V(t)}{\Delta t}. \tag{5.7}$$

If a vector is expressed in terms of its components along the fixed coordinate axes,

$$V = V_1(t)i + V_2(t)j + V_3(t)k, \tag{5.8}$$

there follows

$$\frac{dV}{dt} = \frac{dV_1}{dt} i + \frac{dV_2}{dt} j + \frac{dV_3}{dt} k. \tag{5.9}$$

For the derivative of a product involving two or more vectors the following formulae are used:

$$\frac{d}{dt}(A \cdot B) = A \cdot \frac{dB}{dt} + \frac{dA}{dt} \cdot B \tag{5.10}$$

$$\frac{d}{dt}(A \times B) = A \times \frac{dB}{dt} + \frac{dA}{dt} \times B \tag{5.11}$$

$$\frac{d}{dt}(A \cdot B \times C) = \frac{dA}{dt} \cdot (B \times C) + A \cdot \left(\frac{dB}{dt} \times C\right) + A \cdot \left(B \times \frac{dC}{dt}\right). \tag{5.12}$$

5.8 GEOMETRY OF A SPACE CURVE

Graphically the derivative of a position vector, V(t), defining a space curve is, (see Fig. 5.7) with respect to arc length, s, along the curve, is a unit vector, $\vec{u}$,

$$\left(\vec{u} = \frac{d\vec{r}}{ds} = \frac{dx}{ds} i + \frac{dy}{ds} j + \frac{dz}{ds} k\right) \tag{5.13}$$

tangent to the curve pointing in the direction of increasing arc length.

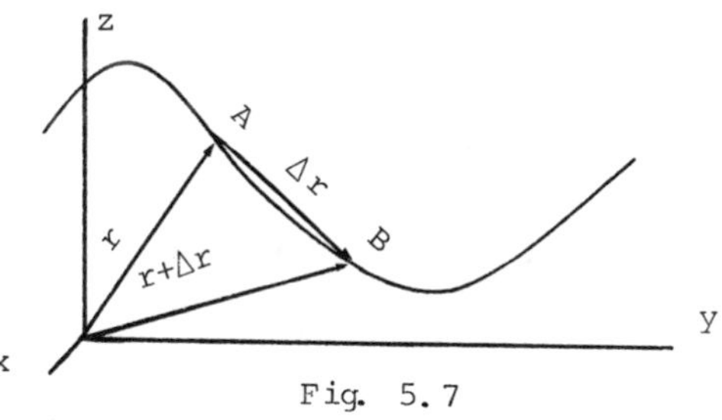

Fig. 5.7

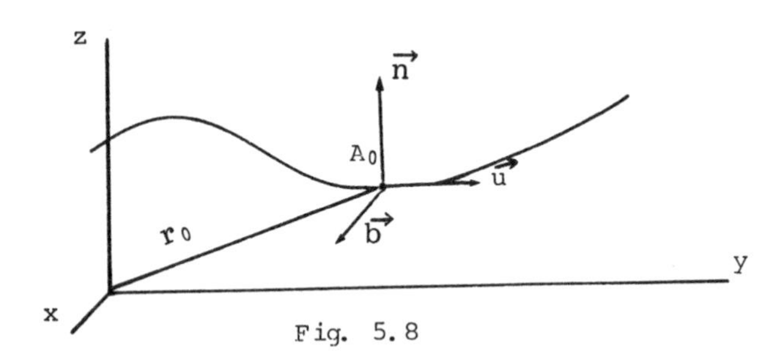

Fig. 5.8

The derivative $\frac{d\vec{u}}{ds} = \frac{d^2\vec{r}}{ds^2} = \frac{d^2x}{ds^2}i + \frac{d^2y}{ds^2}j + \frac{d^2z}{ds^2}k$, of the unit vector $\vec{u}$, is perpendicular to the tangent vector.

$\rho = \frac{1}{\left|\frac{d\vec{u}}{ds}\right|}$ is called the radius of curvature of the space curve. The unit normal vector, or principal normal vector, to $\vec{r}$ is $\vec{n} = \rho\,\frac{d\vec{u}}{ds}$. It points the direction the space curve is "curving."

The binormal vector, defined by $\vec{h} = \vec{u} \times \vec{n}$, is a unit vector perpendicular to the two perpendicular vectors $\vec{u}$ and $\vec{n}$, $\frac{d\vec{b}}{ds}$ is parallel to $\vec{n}$.

τ is the real number which makes

$$\frac{d\vec{b}}{ds} = -\frac{1}{\tau}\vec{n}. \tag{5.14}$$

$\frac{1}{\tau}$ is called the torsion, and the length $|\tau|$ is called the radius of torsion of the curve.

The plane determined at a point A_0 on a curve by $\vec{u}$ and $\vec{n}$ is called the osculating plane; the plane determined by $\vec{n}$ and $\vec{b}$ the normal plane; and that determined by $\vec{u}$ and $\vec{b}$ the rectifying plane (see Fig. 5.8).

The equations of these planes are (5.15)

$\vec{b}_0 \cdot (\vec{r} - \vec{r}_0) = 0,\ \vec{u}_0 \cdot (\vec{r} - \vec{r}_0) = 0,\ \vec{n}_0(\vec{r} - \vec{r}_0) = 0,$

respectively, where $\vec{b}_0$, $\vec{u}_0$, $\vec{n}_0$ are evaluated at $A_0 \in \mathbb{R}^3$, and $\vec{r}_0$ is the position vector to A_0.

5.9 GRADIENT, DIVERGENCE AND CURL

If $\phi(x,y,z)$ a scalar field and $V(x,y,z)$ a vector field have continuous first partial derivatives in a region, we can defined the following:

A) Gradient

$$\text{grad}\ \phi = \nabla\phi = \left(\frac{\partial}{\partial x} i + \frac{\partial}{\partial y} j + \frac{\partial}{\partial z} k\right)\phi$$
$$= \frac{\partial\phi}{\partial x} i + \frac{\partial\phi}{\partial y} j + \frac{\partial\phi}{\partial z} k, \tag{5.16}$$

where $\nabla = \frac{\partial}{\partial x} i + \frac{\partial}{\partial y} j + \frac{\partial}{\partial z} k$ is called the operator del.

B) Divergence

The divergence of V is defined by

$$\text{div}\ V = \nabla \cdot V = \left(\frac{\partial}{\partial x} i + \frac{\partial}{\partial y} j + \frac{\partial}{\partial z} k\right) \cdot (V_1 i + V_2 j + V_3 k)$$
$$= \frac{\partial V_1}{\partial x} + \frac{\partial V_2}{\partial y} + \frac{\partial V_3}{\partial z} \tag{5.17}$$

C) Curl

The curl of V is defined by

$$\text{curl}\ V = \nabla \times V = \left(\frac{\partial}{\partial x} i + \frac{\partial}{\partial y} j + \frac{\partial}{\partial z} k\right) \times (V_1 i + V_2 j + V_3 k)$$
$$= \begin{vmatrix} i & j & k \\ \frac{\partial}{\partial x} & \frac{\partial}{\partial y} & \frac{\partial}{\partial z} \\ V_1 & V_2 & V_3 \end{vmatrix} = i\begin{vmatrix} \frac{\partial}{\partial y} & \frac{\partial}{\partial z} \\ V_2 & V_3 \end{vmatrix} - j\begin{vmatrix} \frac{\partial}{\partial x} & \frac{\partial}{\partial z} \\ V_1 & V_3 \end{vmatrix}$$
$$+ k\begin{vmatrix} \frac{\partial}{\partial x} & \frac{\partial}{\partial y} \\ V_1 & V_2 \end{vmatrix} \tag{5.18}$$

$$= \left(\frac{\partial V_3}{\partial y} - \frac{\partial V_2}{\partial z}\right) i + \left(\frac{\partial V_1}{\partial z} - \frac{\partial V_3}{\partial x}\right) j + \left(\frac{\partial V_2}{\partial x} - \frac{\partial V_1}{\partial y}\right) k.$$

The gradient, divergence and curl in cylindrical and spherical coordinates are shown in the following table:

	Cylindrical coordinates	Spherical coordinates
Conversion to rectangular coordinates	$x = r\cos\phi \quad y = r\sin\phi \quad z = z$	$x = r\cos\phi\sin\theta$ $y = r\sin\phi\sin\theta$ $z = r\cos\theta$
Gradient	$\nabla\phi = \frac{\partial\phi}{\partial r} r + \frac{1}{r}\frac{\partial\phi}{\partial\phi}\phi + \frac{\partial\phi}{\partial z} k$	$\nabla\phi = \frac{\partial\phi}{\partial r} r + \frac{1}{r}\frac{\partial\phi}{\partial\theta}\theta + \frac{1}{r\sin\theta}\frac{\partial\phi}{\partial\phi}\phi$
Divergence	$\nabla \cdot V = \frac{1}{r}\frac{\partial(rV_r)}{\partial r} + \frac{1}{r}\frac{\partial V_\phi}{\partial\phi}$	$\nabla \cdot V = \frac{1}{r^2}\frac{\partial(r^2 V_r)}{\partial r} + \frac{1}{r\sin\theta}\frac{\partial(V_\theta\sin\theta)}{\partial\theta} + \frac{1}{r\sin\theta}\frac{\partial V_\phi}{\partial\phi}$
Curl	$\nabla \times V = \begin{vmatrix} \frac{1}{r} r & \phi & \frac{1}{r} k \\ \frac{\partial}{\partial r} & \frac{\partial}{\partial\phi} & \frac{\partial}{\partial z} \\ V_r & rV_\phi & V_z \end{vmatrix}$	$\nabla \times V = \begin{vmatrix} \frac{r}{r^2\sin\theta} & \frac{\theta}{r\sin\theta} & \frac{\phi}{r} \\ \frac{\partial}{\partial r} & \frac{\partial}{\partial\theta} & \frac{\partial}{\partial\phi} \\ V_r & rV_\theta & rV_\phi\sin\theta \end{vmatrix}$

5.10 DIFFERENTIATION FORMULAS

The following identities are of frequent use:

In these formulas U and V are any vectors and ϕ is a scalar function with all of the required partial derivatives.

$V \cdot \nabla$ is a scalar operator, namely

$$V \cdot \nabla = V_x \frac{\partial}{\partial x} + V_y \frac{\partial}{\partial y} + V_z \frac{\partial}{\partial z}$$

$\nabla \cdot (\phi V) = \phi(\nabla \cdot V) + V \cdot (\nabla \phi)$

$\nabla \times (\phi V) = \phi(\nabla \times V) + (\nabla \phi) \times V$

$\nabla \cdot (V \times U) = U \cdot (\nabla \times V) - V \cdot (\nabla \times U)$

$\nabla \times (V \times U) = U \cdot \nabla V - V \cdot \nabla U + V(\nabla \cdot U) - U(\nabla \cdot V)$

$\nabla (V \cdot U) = V \cdot \nabla U + U \cdot \nabla V + V \times (\nabla \times U) + U \times (\nabla \times V)$

$\nabla \times (\nabla \phi) = \text{curl grad}\, \phi = 0$

$\nabla \cdot (\nabla \times V) = \text{div curl } V = 0$

$\nabla (\phi_1 + \phi_2) = \nabla \phi_1 + \nabla \phi_2$, ϕ_1 and ϕ_2 scalar functions with partial derivatives

$\nabla \cdot (V + U) = \nabla \cdot V + \nabla \cdot U$

$\nabla \times (V + U) = \nabla \times V + \nabla \times U$

$\nabla \cdot (\nabla V) = \nabla^2 V = \frac{\partial^2 V}{\partial x^2} + \frac{\partial^2 V}{\partial y^2} + \frac{\partial^2 V}{\partial z^2}$ is called the Laplacian of V, and

$$\nabla^2 \equiv \frac{\partial^2}{\partial x^2} + \frac{\partial^2}{\partial y^2} + \frac{\partial^2}{\partial z^2} \qquad (5.19)$$

is called the Laplacian operator.

Note that $\nabla^2 u$ is not defined for a vector U.

5.11 THE POTENTIAL FUNCTION

Let $\vec{V}(\vec{x})$ be a vector field defined on an open set U in n-dimensional space. That is, for each vector $\vec{x} = (x_1, x_2, \ldots, x_n)$ which represents a point $P = (x_1, x_2, \ldots, x_n)$ in n-space, we associate the vector $\vec{F}(\vec{x})$ of dimension m.

Then if $\phi(x)$ is a differentiable real valued function on U such that $\vec{F} = \nabla \phi$, it is said that ϕ is a potential function for $\vec{F}$.

If $\vec{F}$ represents force, the negative of $\phi(\vec{x})$ is called the potential energy associated with $\vec{F}$. When such a function $\phi(\vec{x})$ exists, and is single-valued in a region, the force $\vec{F}$ is said to be conservative.

5.12 DIRECTIONAL DERIVATIVES

If the real-valued function $f(x,y,z)$ is defined at a point $P(x,y,z)$ on a given space curve and $f(x + \Delta x, y + \Delta y, z + \Delta z)$ is the value of the function at a neighboring point on C at a distance Δs from the point $P(x,y,z)$, where s is the distance function along C, then the directional derivative of f at the point $P(x,y,z)$ along the curve C is given by

$$\frac{df}{ds} = \lim_{\Delta s \to 0} \frac{\Delta f}{\Delta s} = \lim \frac{f(x+\Delta x, y+\Delta y, z+\Delta z)-f(x,y,z)}{\Delta s} \tag{5.20}$$

$$= \frac{\partial f}{\partial x}\frac{dx}{ds} + \frac{\partial f}{\partial y}\frac{dy}{ds} + \frac{\partial f}{\partial z}\frac{dz}{ds}.$$

In vector form this can be written

$$\frac{dF}{ds} = \left(\frac{\partial F}{\partial x} i + \frac{\partial F}{\partial y} j + \frac{\partial F}{\partial z} k\right) \cdot \left(\frac{dx}{ds} i + \frac{dy}{ds} j + \frac{dz}{ds} k\right) \tag{5.21}$$

$$= (\nabla f) \cdot \frac{dr}{ds} = (\nabla f) \cdot T$$

where T is the unit tangent vector to C at (x,y,z).

5.13 APPLICATIONS OF PARTIAL DERIVATIVES TO GEOMETRY

If $f(x,y,z) = 0$ is the equation of a surface S, then the following equations can be defined:

A) Tangent Plane to a Surface at P

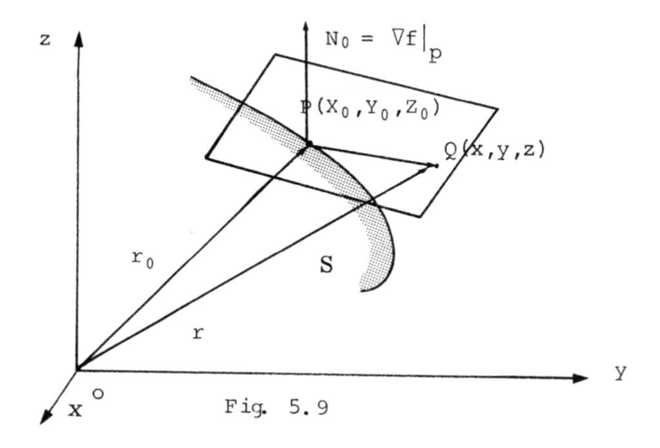

Fig. 5.9

The equation, in variable r, of the plane is

$$(\vec{r} - \vec{r}_0) \cdot N_0 = (\vec{r} - \vec{r}_0) \cdot \nabla f\Big|_P = 0 \tag{5.22}$$

where r_0 is a fixed vector drawn from the origin to P. In rectangular form the equation is:

$$\left.\frac{\partial f}{\partial x}\right|_P (x - x_0) + \left.\frac{\partial f}{\partial x}\right|_P (y - y_0) + \left.\frac{\partial f}{\partial z}\right|_P (z - z_0) = 0 \tag{5.23}$$

B) Normal Line to a Surface

The equation for the normal line to a surface is

$$(\vec{r} - \vec{r}_0) \times \vec{N}_0 = (\vec{r} - \vec{r}_0) \times \nabla f\Big|_P = 0. \tag{5.24}$$

In rectangular form this is

$$\frac{x - x_0}{\left.\frac{\partial f}{\partial x}\right|_P} = \frac{y - y_0}{\left.\frac{\partial f}{\partial y}\right|_P} = \frac{z - z_0}{\left.\frac{\partial f}{\partial z}\right|_P}. \tag{5.25}$$

C) Tangent Line to a Curve

The tangent line to the curve C parameterized by $\vec{R}(t)$ is given by the equation

$$(\vec{r} - \vec{r}_0) \times \vec{T}_0 = (\vec{r} - \vec{r}_0) \times \left.\frac{dR}{dt}\right|_P = 0, \tag{5.26}$$

where

$$\vec{R} = f(t)i + g(t)j + h(t)k$$

f,g,h are defined as continuously differentiable functions

$$T_0 = \left.\frac{dR}{dt}\right|_P.$$

In rectangular form equation (5.26) is

$$\frac{x - x_0}{f'(t_0)} = \frac{y - y_0}{g'(t_0)} = \frac{z - z_0}{h'(t_0)}. \qquad (5.27)$$

If the curve C is given as the interrection of the surfaces $S_1(x,y,z) = 0$ and $S_2(x,y,z) = 0$, the correspnoding equation of the tangent line is

$$\frac{x - x_0}{\left|\begin{matrix} S_{1y} & S_{1z} \\ S_{2y} & S_{2z} \end{matrix}\right|_P} = \frac{y - y_0}{\left|\begin{matrix} S_{1z} & S_{1x} \\ S_{2z} & S_{2x} \end{matrix}\right|_P} = \frac{z - z_0}{\left|\begin{matrix} S_{1x} & S_{1y} \\ S_{2x} & S_{2y} \end{matrix}\right|_P}. \qquad (5.28)$$

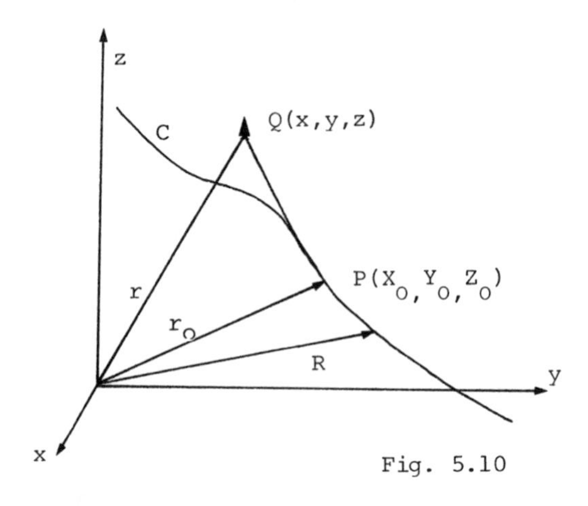

Fig. 5.10

D) Normal Plane to a Curve

The equation of the normal plane to a curve is

$$(\vec{r} - \vec{r}_0) \cdot \vec{T}_0 = (\vec{r} - \vec{r}_0) \cdot \left.\frac{dR}{dt}\right|_P = 0. \qquad (5.29)$$

In rectangular form equation (5.29) is

$$f'(t_0)(x - x_0) + g'(t_0)(y - y_0) + h'(t_0)(z - z_0) = 0. \qquad (5.30)$$

When the curve is defined by two surfaces S_1 and S_2, then

$$\begin{vmatrix} S_{1y} & S_{1z} \\ S_{2y} & S_{2z} \end{vmatrix}_P (x - x_0) + \begin{vmatrix} S_{1z} & S_{1x} \\ S_{2z} & S_{2x} \end{vmatrix}_P (y - y_0)$$

$$+ \begin{vmatrix} S_{1x} & S_{1y} \\ S_{2x} & S_{2y} \end{vmatrix}_P (z - z_0) = 0. \qquad (5.31)$$